INNOVATIONS IN THE MODERN ARTS AND SCIENCES

Explaining Lasting Change in Novelty-Driven Cultural Fields

By
Richard Whitley and Jochen Gläser

BRISTOL
UNIVERSITY
PRESS

First published in Great Britain in 2026 by

Bristol University Press
University of Bristol
1–9 Old Park Hill
Bristol
BS2 8BB
UK
t: +44 (0)117 374 6645
e: bup-info@bristol.ac.uk

Details of international sales and distribution partners are available at bristoluniversitypress.co.uk

DOI: 10.51952/9781529239393

British Library Cataloguing in Publication Data
A catalogue record for this book is available from the British Library

ISBN 978-1-5292-3937-9 paperback
ISBN 978-1-5292-3938-6 ePub
ISBN 978-1-5292-3939-3 OA PDF

Cover design: Newgen
Front cover image: iStock/akinbostanci
Bristol University Press uses environmentally responsible print partners.
Printed and bound in Great Britain by CPI Group (UK) Ltd, Croydon, CR0 4YY

FSC
www.fsc.org
MIX
Paper | Supporting responsible forestry
FSC® C013604

Bristol University Press' authorized representative in the European Union is:
Easy Access System Europe, Mustamäe tee 50, 10621 Tallinn, Estonia,
Email: gpsr.requests@easproject.com

For Barbara and Grit

Contents

List of Tables

About the Authors

Richard Whitley is Emeritus Professor of Organisational Sociology at the University of Manchester. He has undertaken extensive work in the sociology of science and has also been a founding influence in the comparative study of societal business systems. In 2007 he was elected a Foreign Member of the Royal Swedish Academy of Letters, History and Antiquities. Richard's publications include: *The Intellectual and Social Organisation of the Sciences* (Oxford University Press, 1984, 2000), *Business Systems in East Asia* (Sage, 1992), *Divergent Capitalisms: The Social Structuring and Change of Business Systems* (Oxford University Press, 1999), *Business Systems and Organisational Capabilities: The Institutional Structuring of Competitive Competences* (Oxford University Press, 2007), *The Changing Governance of the Sciences* (Springer, 2007) and *Organizational Transformation and Scientific Change* (Emerald, 2014).

Jochen Gläser is Professor of Social Studies of Science and Technology at the Technical University Berlin. His major research interests include the connection between epistemic practices and social structures of scientific communities, the interaction between academic careers and knowledge production, and the methodology of social studies of science. His publications include *Wissenschaftliche Produktionsgemeinschaften: Die soziale Ordnung der Forschung* (Campus, 2006*), Changing Funding Arrangements and the Production of Scientific Knowledge* (Special Issue of *Minerva*, 2018) and 'From Effects of Governance to Causes of Epistemic Change' in *Minerva* (2024).

Preface

The modern arts and sciences are fields of cultural production in which the creation of original works has become an everyday activity generating a large variety of novel products. However, only some of these outputs become innovations that have a lasting effect on the kinds of problems that other competent producers work on and how they do so. It is the extent to which novel ideas, techniques and results alter the nature, organization and valuation of artistic and scientific activities that determines their long-term importance as aesthetic and epistemic innovations.

An important question in the analysis of changing practices and goals in the modern arts and sciences, then, is how, and under what conditions, do particular kinds of innovations become established and alter dominant ways of doing things in different fields? These processes can be expected to vary between differently organized and governed artistic and scientific fields as well as being affected by the nature of the broader socio-economic circumstances in which they are conducted.

In this book we develop an approach for dealing with this question by presenting an explanatory framework for analysing how different kinds of lasting change are developed and institutionalized in different types of artistic and scientific fields. By building on a number of case studies from the sociology of science, history of science, history of the arts and cultural economics, we show how the development of different innovations are affected by dominant patterns of coordination and control and demonstrate that it is essential to take differences in these features into account when analysing aesthetic and epistemic changes.

This comparative framework helps us understand how the changes in socio-economic conditions have been re-shaping the processes by which innovations in the arts and sciences produce lasting change by modifying the productive relationships between practitioners in the arts and sciences and the nature of authority relations in art worlds and science worlds.

Earlier versions of these ideas and case studies have been presented to conferences of the European Sociological Association (2019), the British Society for Literature and Science (2022), and workshops held at the Universities of Rotterdam (2021) and Goethe Universität, Frankfurt (2023).

We are very grateful to participants at these meeting for their comments, questions and suggestions.

We are also greatly indebted to Grit Laudel, Uwe Schimank and Friedrich Steinle for their thorough reading of the first draft of the typescript and helpful suggestions and to Andy Wilson for his help with the analysis of the Bebop innovation.

Richard Whitley and Jochen Gläser
January 2026

1

Introduction

Understanding change in the arts and sciences

A central problem in accounting for artistic and scientific change is to understand how the everyday production of novelty in these fields of cultural production can result in consequential shifts in practitioners' practices, goals and evaluations. Under what conditions, by what processes and with what effects do particular contributions alter the kinds of problems that artists and scientists work on, how they do so and how they judge results? As Richard Caves (2000, pp 2–4) emphasized in his analysis of the creative industries in the late 20th-century USA, a constitutive feature of systems of creative practices is their habitual production of an infinite variety of novel outputs. These products vary greatly in their collectively perceived relevance and impact on future creations, with only some of them triggering innovations that change practitioners' practices, goals and evaluations on a continuing basis.

Unlike the technical product innovations that have been the prevalent focus of much innovation research, innovations in novelty-producing fields are only the 'visible tip of the iceberg of everyday creativity' (Caves, 2000, p 202). Novelties become innovations when they are recognized as significant contributions that shape at least some of the subsequent production of novelties and so generate lasting change in practices and purposes. As Howard Becker suggested for the arts: 'Innovations begin as, and continue to incorporate, changes in an artistic vision or idea. But their success depends on the degree to which their proponents can mobilize the support of others. Ideas and visions are important, but their success and permanence rest on organization, not on their intrinsic worth' (Becker, 1982, pp 309–10).

While some major changes in the modern arts and sciences have been studied by historians and sociologists, the problem of explaining the processes by which only a limited number of new outputs generate such changes in fields whose everyday activities require the production of novelties has rarely been explicitly addressed. Most studies of lasting change are either highly detailed case studies of particular changes, which do not directly contribute to the comparative analysis of commonalities and differences between cases, or highly abstract models (primarily in science studies) that tend to ignore variations in socio-economic conditions and the processes leading to lasting change. Neither type of study can answer the question of how different kinds of lasting change occur in differently organized fields of cultural production that constantly generate novelties.

In considering processes of lasting change in the arts, it is important to recognize that the modern system of the fine arts displaced the previously dominant patronage model of artistic/artisanal production and evaluation in Europe over the course of the long 18th century (1680–1830). It established new ideas of the nature of art works, of the identity and role of artists and of the appropriate standards for evaluating artistic creations, as well as new institutions such as museums, public concerts and copyright (Shiner, 2001, pp 10–5, 126–9). Studies of lasting change in the arts since this modern system became established have tended to focus on relatively radical and far-reaching shifts in aesthetic practices, purposes and organizational structures, such as the rapid development and diffusion of Cubism in painting and sculpture (Golding, 1968; Cottington, 1998, 2004, 2022; Cox, 2000; Sgourev, 2013) or the rise of what White and White (1965) termed the dealer-critic system governing the production and dissemination of new paintings and sculptures in France. These studies have mostly concentrated on the detailed reconstructions of individual cases in particular circumstances rather than exploring how their patterns of development vary between different contexts. The lack of shared frameworks for characterizing and explaining variations in these patterns has made it difficult to undertake systematic comparisons.

Additionally, these case studies illustrate two difficulties in comparing processes of lasting change across artistic genres and national contexts. First, they often take place in different ways and circumstances in literature, painting, music and other genres. As Shiner (2001, pp 11–2) has pointed out, during the development of the modern art system the prestige and role of the poet changed quite differently to that of the painter, and the separation of imaginative literature from literature in general did not coincide with the split of the visual arts into fine art versus craft. Similarly, musicians remained dependent on patronage for support longer than did writers and painters, and performing for pay was commonly regarded as a low status occupation for much of the 18th century (Goehr, 1992, pp 176–82; Shiner, 2001, pp 108–11).

Second, changes of socioeconomic conditions, of the nature of audiences, of organizational structures of artistic fields and of other important conditions for lasting artistic change, developed quite unevenly across countries. For example, painting, sculpture and architecture and their makers achieved liberal arts status in Italy well before they did in the rest of Europe, and the institutionalization of authorial copyright laws across Europe was both faster for literary works than for musical compositions and occurred earlier for literature in Britain than in France and Germany (Goehr, 1992, pp 218–20; Woodmansee, 1994; Bonds, 1996). These differences have often restricted the study of lasting change in a particular field to one country.

Most accounts of lasting changes in the sciences have tended either to outline highly general models of major transformations, often derived from examples in a small number of fields, or to provide detailed descriptions of changes in practices and goals within particular areas, such as case studies of the emergence of new scientific fields. The most influential account of lasting change in the sciences has been Thomas Kuhn's (1970 [1962]) model of paradigm replacement in 'mature' sciences, which he understood as those where research is guided by a 'paradigm'.

In these sciences researchers solve 'puzzles', that is problems set by the paradigm with approaches provided by the paradigm. In the course of this puzzle solving 'anomalies' – results that pose problems that should be soluble but in fact are insoluble in the terms of that paradigm (Masterman, 1970, p 82) – occur and need to be dealt with. As the number of anomalies that cannot be resolved within the framework of the current paradigm increases, the paradigm weakens, and a 'crisis' occurs which leads to the old paradigm being replaced with a new one that can integrate the problems that were anomalies to the previous one. The latter case, which Kuhn terms a 'scientific revolution', represents a lasting change of practices of knowledge production.

Kuhn's model has been highly influential in the philosophy, history and sociology of science. It triggered investigations of the relationship between epistemic and social structures in scientific communities and research on the emergence of research specialties. It also helped legitimize empirical research in the social construction of scientific knowledge, which largely displaced Mertonian institutionalism as the mainstream of the sociology of science.

In spite of its indisputable influence in several areas of science studies, Kuhn's account cannot be used to explain how different forms of lasting change take place in particular circumstances. This is largely due to the parsimony of his model, which specifies just one epistemic condition that triggers change (the accumulation of anomalies), one epistemic process of change (the replacement of a paradigm) and one outcome (normal science guided by a new paradigm). The model is internalist in taking epistemic change as its object without referring to anything other than itself in two respects.

First, it ignores the social forms in which epistemic change occurs, that is the structures of scientific communities, the practices of knowledge production and the social processes that bring about their change. Second, it ignores the changing socio-economic conditions under which the sciences produce knowledge and which produce varying levels of collective practitioner community autonomy, societal expectations and access to resources. Without taking social processes and socio-economic conditions into account, it is impossible to explain how processes of lasting change take place and vary with the circumstances under which they occur. The same applies to alternatives to the Kuhnian model that have been proposed by philosophers of science, such as Lakatos' model of competing research programmes (Lakatos, 1968).

The explanatory power of Kuhn's model is also limited by its narrow scope in focusing on only one type of research and only one process of change. The model of paradigm replacement applies only to 'mature' sciences in which research is guided by a paradigm. The concept of maturity here implies that all fields of research mature by converging towards the model of paradigm-guided research, an assumption that seems doubtful even for the natural sciences in light of their diversity (see Galison and Stump, 1996, on the 'disunity of science').

The varied nature of forms of knowledge production is further exemplified by the social sciences and humanities, which are not immature fields that take more time to develop into Kuhnian 'mature' sciences but rather are established fields that create second-order constructs. As Schütz (1962, p 59) emphasized, 'the constructs of the social sciences are … constructs of the second degree, that is, constructs of the constructs made by the actors on the social scene, whose behavior the social scientist has to observe and to explain in accordance with the procedural rules of his science'. It seems unlikely that such fields will fit a model of knowledge production whose template was derived from the physical sciences. Kuhn did not answer the question how lasting change occurs in non-paradigmatic or multi-paradigmatic fields.

The scope of Kuhn's model is further limited by its inclusion of only one process of change, the exchange of one paradigm for another. Such a complete replacement is unlikely because practices of knowledge production are highly diverse and vary in their epistemic and practical integration and cohesion, and some of them are likely to persist even in the light of paradigm change. Furthermore, subsequent empirical research soon produced examples of lasting change that did not fit Kuhn's model. Among these, the emergence of new fields of inquiry, described by Mulkay (1975) as 'branching', became most prominent. A wave of case studies on the emergence of scientific fields underscored this point. However, since these were case studies of individual fields that derived their interpretations from singular reconstructions, the attempt by Edge and Mulkay (1976, pp 350–402) to compare them systematically yielded few theoretical insights.

Consequently, neither the abstract models of radical conceptual change nor empirical studies of the emergence of new scientific fields have led to a systematic account of the processes that bring about lasting change of scientific practices or the conditions under which these processes occur. The same can be said for the historiographical accounts of scientific revolutions (for example, Cohen, 1985), great discoveries (Pinch, 1986) or scientific 'breakthroughs' (for example, Hollingsworth, 2008; Heinze et al, 2009). These studies mostly focus on the conditions under which exceptional scientific findings are produced and neglect the conditions under which they make fellow researchers change their practices. Micro-sociological studies of lasting change in scientific practices contributed rich material on changes in practices and the conditions under which they occurred but again were single-case studies that did not lead to broader conclusions about the circumstances in which particular kinds of changes are more or less likely to come about (Fujimura, 1988; Cambrosio and Keating, 1995).

This near-complete absence of comparative perspectives on the processes that lead from particular scientific findings to lasting changes of scientific practices is mostly due to the large conceptual distance between highly abstract models and detailed historical or sociological reconstructions of individual cases. In addition, studies of scientific change face major conceptual and methodological problems derived from the complexity of empirical investigations that trace changes in practices in international scientific communities while taking into account the variations in both national conditions and the epistemic properties of research practices in different fields (Whitley, 2014; Whitley et al, 2018).

A major obstacle to progress in accounting for lasting changes of artistic or scientific practices appears to be the diverse nature of these practices in different arts and sciences. It seems very difficult to construct an account that is applicable to all fields and is specific enough to have explanatory power. Not surprisingly, comparisons of processes of change across both the arts and the sciences have been even rarer, partly because of the common view that creating art works and researching scientific problems are quite different kinds of cultural production activities.

One of the few exceptions is the attempt by Clignet (1985) to apply Kuhn's model of paradigm change to the arts. Although this use of Kuhn's model required a high level of abstraction, Clignet's comparison highlights the narrow scope of Kuhn's model. He observed that partial paradigm change amidst continuation of earlier styles and goals is more common in many artistic fields than radical transformations of dominant practices and conventions, not least because of the importance of non-practitioner audiences in recognizing and evaluating innovations (Clignet, 1985, p 74; see also Becker, 1982, 304–10). Although Kuhn did not discuss the relative frequency of transformational scientific revolutions that followed his model, subsequent research suggests that they are rare in the sciences, too. Since Clignet argued at the same high level of abstraction as Kuhn, he was similarly unable to account for the conditions under which specific kinds of 'artistic revolutions' were successful or for the kinds of processes by which this success was achieved.

In his extensive analyses of artistic and scientific fields as social structures of domination between groups and individuals that vary in their autonomy from broader structures of domination, Bourdieu (1975, 1993, 2004, 2017) emphasized the common nature of struggles between artists and scientists in these fields for authority over aesthetic and epistemic standards and access to resources. Authority relationships in both sets of fields structure what is possible for actors in different positions to work on and achieve (Bourdieu, 1993, pp 67–73). They also function as intermediaries between general socio-political changes, such as the expansion of painters competing for training and recognition in 19th-century France and the growth of middle class demand for genre works that had low status in the formal Academy hierarchy, and related shifts in authority between the dominant Academy training and consecration system and those groups challenging it, as in the case of Manet's symbolic revolution (Bourdieu, 2017, pp 245–69). However, it remains unclear how different patterns of aesthetic and epistemic change are more or less likely to occur and establish major innovations in differently organized fields of cultural production and socio-economic conditions or, indeed, how practitioners in different artistic and scientific fields achieve autonomous control over aesthetic and epistemic standards and resource allocation procedures.

The crucial role of producer communities in the generation and use of innovations in all fields of cultural production has been highlighted by Diana Crane (1972, pp 129–42) in her call for an integrated sociology of culture. She later suggested a framework for exploring how differences in producers' access to symbolic and material rewards in artistic, scientific and religious fields are likely to lead to differences in the variety and continuity of 'innovations' between fields, emphasizing the commonality of the sciences, art and religion as fields of cultural production (Crane, 1976). However, since she did not distinguish between novelties in general and innovations (she terms all novelties 'innovation'), and since she compares only one condition (the reward systems), her comparative approach does not provide an explanation of how some novelties become innovations producing lasting change.

Although the approaches by Clignet, Bourdieu and Crane do not provide explanations of the processes by which lasting change is achieved, they imply that the modern arts and sciences are similar fields of cultural production in their common institutionalization of distinct collectivities of artists and scientists that produce, evaluate and use novel contributions to collective purposes. These collectivities do vary considerably, though, in the extent of their organizational integration, autonomy and form of governance across fields and socio-economic contexts.

An important feature of these producer communities in both the arts and the sciences is that the recognition of practitioners as members depends on their continued production of novel outputs that are considered relevant to the community's collective goals, albeit to varying degrees of significance for, and impact on, others' projects. Lasting change in these kinds of collectively maintained activities depends on everyday novelty production but is only achieved when novel outputs influence other competent producers' practices and purposes. Crucial to the establishment of innovations that change how artists and scientists work and on what problems in different fields, then, are the processes by which novelties become innovations.

These common organizational features of the modern arts and sciences, and their varying nature between fields and contexts over time, enable us to compare the processes through which different kinds of lasting aesthetic and epistemic changes, or innovations, become established in differently organized fields. They also facilitate the analysis of how changing socio-economic conditions can be expected to affect these processes and so the likelihood of significant changes in dominant practices, problems and outputs in different fields. An important question in the analysis of changing problems and approaches in the modern arts and sciences, then, is how, and under what conditions, do particular kinds of innovations become established and alter dominant ways of doing things in differently organized fields and changing contexts?

Developing and establishing innovations in the modern arts and sciences

In this book we develop an approach for dealing with this question by presenting an explanatory framework for analysing how different kinds of innovations are developed and institutionalized in different types of artistic and scientific fields, drawing on empirical research from the sociology of science, history of science, history of the arts and cultural economics. Applying this framework to nine reconstructions of the development of different kinds of innovations in different conditions, we show how they were affected by variations in both the distribution of aesthetic/epistemic authority and control over resource allocation in each field and the nature of their socio-economic contexts. We demonstrate that it is essential to take differences in these features into account when analysing aesthetic and epistemic changes. This comparative framework helps us to understand how changes in socio-economic conditions have been re-shaping the processes through which innovations in the arts and sciences produce lasting change by modifying the productive relationships between practitioners in the arts and sciences and the nature of authority relations in art worlds and science worlds. The structure of our argument can be summarized as follows.

In the next chapter we outline how the modern arts and sciences have become institutionalized as social systems of cultural production that organize the generation of novel contributions and competition for legitimacy and resources in comparable ways as distinct novelty-driven fields. These are fields in which novelty production is not only institutionalized as the major purpose of creative practices but also influences subsequent production practices by shaping their opportunity structure as input, counterpoint, context or other point of reference to which practitioners respond. Although novelty production practices and goals differ between the arts and the sciences – with the former being primarily concerned with the creation and evocation of imaginary worlds and the latter focused on the advancement of intersubjectively robust propositional knowledge about shared objects of inquiry (Meyer, 1974; Berger, 2000, pp 57–63) – in both areas of cultural production practitioners compete for the collective recognition of their products as legitimate and significant novelties for collective aesthetic/epistemic purposes, and need to respond to each other's novelty production in this competition.

Understanding the modern arts and sciences as novelty-driven fields requires reconstructing their institutionalization as such for three reasons. First, the conditions facilitating or hindering the development of different kinds of innovations in different arts and sciences frequently change, which makes any systematic account of how innovations produce aesthetic and epistemic change applicable to a particular historical periods of their development. The historical scope of this account can be established by identifying the period in which these conditions were in place and remained sufficiently stable. This includes considering the consequences of further changes of these conditions, which will be briefly discussed in Chapter 6. Second, reconstructing the institutionalization of social structures in which novelty-driven production has been taking place contributes to a better understanding of their dynamics in the current period, whether as residuary from previous developments or endemic to the modern arts and sciences. Finally, the conditions for innovations in the modern arts and sciences do in turn depend on the dynamics of wider social contexts, and the reconstruction of the institutionalization of the modern arts and sciences contributes to understanding the link between them and societal development. These processes can be understood by extending Becker's (1982) concept of art worlds to the sciences and distinguishing types of art worlds and science worlds according to their organizational features.

The processes by which some novelties lead to different kinds of lasting change in the modern arts and sciences are discussed in Chapter 3. For novelties to become innovations, they have to be publicly available, accepted as legitimate and alter the nature of fellow practitioners' practices and purposes on a continuing basis (Galenson, 2006, p 8; North, 2013, p 4).

Unlike the Schumpeterian sharp separation of invention, innovation and imitation processes in many studies of technological change (Rosenberg, 1975), we adopt a developmental understanding of innovations in the modern arts and sciences in which some novelties become 'inventions' which are collectively developed by practitioners in 'innovation' processes of subsequent continuously changing novelty production. It is the extent to which these processes affect the characteristics, organization, and valuation of subsequent artistic and scientific activities and outputs that determines their long-term importance as aesthetic and epistemic innovations.

Such innovation processes vary in the way in which they change novelty production. In particular, they differ in the extent to which they alter established patterns of working, dominant objectives and evaluation standards and in the degree to which these changes become obligatory for all competent producers.

Drift innovations consist of minor changes in how work is conducted and for what purposes, such as shifts in performing practices in the arts or technical refinements in research. **Process innovations** establish more distinct and discontinuous methods and approaches for dealing with common problems. **Complementary innovations** establish the pursuit of new priorities and aesthetic or epistemic problems, which usually involves more substantial deviations from established norms and practices, but without requiring the wholesale adoption of these new approaches throughout the field. **Displacement innovations** replace the dominant priorities and aesthetic or epistemic problems, thereby significantly altering the nature of skills required of competent practitioners and prevalent conceptions of the field. In contrast, *fads* are short-term enthusiasms that have little or no consequential impact on the nature of later collective practices, problems and standards.

In Chapter 4 we distinguish types of art worlds and science worlds in terms of two organizational characteristics that structure incentives and opportunities to change practices of novelty production: (a) the dominant mode of productive relationships between practitioners in the generation of novelties, and (b) the distribution of aesthetic/epistemic and resource allocation authority between practitioners, intermediaries and audiences. The combination of different kinds of productive relationships between practitioners with different distributions of authority over standards in each artistic and scientific field enables us to distinguish between three major kinds of art worlds and science worlds.

Practitioners in *intermediary-led* art worlds and science worlds have relatively weak levels of mutual dependence with each other's outputs in producing novelties and exercise limited collective authority over the production of new works and the standards for evaluating them. In *practitioner-led* art worlds and science worlds, practitioners have closer productive connections with each other's outputs and are able to exert considerable aesthetic and epistemic authority over standards and priorities, but have to share such authority with well-informed audiences and intermediaries. In *practitioner-dominated* science worlds, practitioners combine close productive relationships with each other's outputs with high levels of collective authority over epistemic priorities and practices.

How different kinds of innovations are developed and become established in these differently structured art worlds and science worlds is explored in nine case studies of successful innovations in Chapter 5. The ability to develop innovations in these worlds depends on the availability of opportunities for practitioners to build protected space in which novel problems and goals can be pursued without suffering severe reputational and career consequences, and the distribution of authority between competent practitioners, audiences and intermediaries. The presentation of each case includes descriptions of the nature of the innovation, its development in practitioner communities and the conditions facilitating this development.

The two innovations developed in intermediary-led art worlds, Bebop and the *Nouvelle Vague* of film making in French cinema, showed how the combination of the weakening authority of major intermediaries, shifts in some of their priorities, changes in the nature of audiences and in the distribution of outputs to them enabled ambitious artists to develop new styles and establish displacement and complementary innovations respectively. In the four cases of innovations developed in practitioner-led art and science worlds, Cubism, Metafiction, Oncogene Approach and Financial Economics, changes in the nature, roles, interests and authority of both intermediaries and practitioners led to the development of displacement and complementary innovations in art, literature, cancer research and business finance research. In the last area, the transformation of the polycentric field of business finance into the more integrated field of Financial Economics was supported by pressure from philanthropic foundations and other intermediaries and audiences for business schools to become more research-based and the attraction of the assumptions and methods of neo-classical economics to practitioners seeking intellectual acknowledgement of their contributions.

In the two cases of major process innovations in practitioner-dominated integrated scientific fields, Radio Astronomy and Bose-Einstein Condensation (BEC), the availability of substantial and continued public financial support for new instruments and personnel was crucial to their successful development, as was the general recognition of the wider scientific community that they were worthy of such support. Finally, the development of Evolutionary-Developmental Biology (Evo-devo) as a complementary innovation was assisted by the widespread recognition by the scientific audience of the inadequacies of neo-Darwinian approaches to evolutionary biology in dealing with recent empirical findings together with the contemporary development of molecular-biological methods for investigating developmental processes in different model organisms. As with Radio Astronomy and BEC, Evo-devo also benefitted from intermediary support for this kind of long term, fundamental research as well as employers' continued willingness to provide suitable infrastructural support.

These nine case studies of successful innovations in differently organized art and science worlds have shown how the necessary protected space for their development and acceptance depended on a variety of combinations of practitioner actions, shifts in intermediary roles, and priorities and changes in the nature and interests of audiences. In the first part of Chapter 6 we discuss the nature of the particular combinations of changes in intermediary-led art worlds, practitioner-led art and science worlds, and practitioner-dominated science worlds that provided the protected space for practitioners to develop and gain acceptance for these innovations and consider their broader implications for understanding how changes in the modern arts and sciences take place.

In the second part of this chapter, we discuss two of the significant changes in the socio-economic conditions governing authority relationships in many art and science worlds. First, the rapid expansion of transnational markets in contemporary art works, associated intermediary roles and agents across the world that led to an emerging bifurcation of aesthetic goals and standards between those governed by a logic of artistic prestige and charisma and those dominated by a commercial logic. Second, the expansion of state policies and agencies evaluating and guiding the use of public funds and associated forms of support that have shifted authority relationships in many art and science worlds.

The book concludes with the third part of the chapter, which summarizes the main points being developed in this book: the historical contingency of novelty-driven production, an understanding of innovations as processes of continuous lasting change in the production of novelties, and the possibility to achieve theoretical accounts of conditions, causes and consequences of such change by comparatively studying innovations. Although our main contribution is conceptual, the comparative analysis of published case studies demonstrates that it is possible to compare processes bringing about lasting change in the arts and sciences, and to use these comparisons for developing theories of lasting change in fields of cultural production.

2

The Institutionalization of the Modern Arts and Sciences as Novelty-Driven Fields

Novelty-driven fields of cultural production

The comparative framework applied in this book is based on the premise that the modern arts and sciences share a mode of production that can be characterized as novelty-driven. Novelty-driven cultural production occurs when the production of novelties is not only institutionalized as the major purpose of creative practices but also contributes to shaping subsequent production by providing inputs, counterpoints, contexts or other points of reference to which practitioners respond if they want to make a relevant contribution. These responses occur in the context of practitioners' mutual awareness and perception of addressing similar aesthetic or epistemic problems. The ways in which such awareness of, and responses to, others' contributions affect practitioners' subsequent production of novelties vary considerably between historical periods and different arts and sciences. In this chapter we provide a reconstruction of the institutionalization of novelty driven production in the arts and sciences and outline the social structures of modern art worlds and science worlds that developed in this institutionalization process.

Creative responses by practitioners to previously produced novelties involve their quite autonomous interpretation and use of predecessors' contributions for their own productive purposes rather than reflecting the latter's influence, as Baxandall (1985, pp 58–62) has emphasized for the visual arts. A clear indication of autonomous responses to previous novelty is the use of novelties produced by creative practitioners who are long deceased. In painting, Fry has illustrated this relationship for the case of Cubism: 'the special achievement of Cubism, and above all of Picasso, was to reinvent classical, mediated representation and in that reinvention also to transform it so as to reveal its central conventions and mental processes' (Fry, 1988, p 296).[1] According to Fry, Picasso was not so much influenced by the classics as actively reinterpreting and using them as material for his own production of novelties (see also Baxandall, 1985, p 61).

The notion of relevant contributions implies that the community's collective identity includes a shared understanding of the nature of the field's constitutive aesthetic/epistemic problems that are jointly or individually addressed by members, such as those involved in the development of Impressionism, Post-impressionism and Cubism in late 19th and early 20th-century French painting (Baxandall, 1985, pp 44–5). It also implies some collective authority of practitioner communities over the recognition of members' works as legitimate, which in turn requires some collective producer autonomy from other actors and communities. Such communities matter for the recognition of novelty because their standards and judgments, be they implicit or explicit, are not dependent on the approval of intermediaries or audiences.[2] Novelty-driven fields of cultural production are thus centred on a community of creative practitioners whose collective identity is based on practitioners' awareness of shared problems and ideas of what counts as a relevant contribution. Members of these communities respond to each other's production, use novelties produced by some of their fellow community members in their own production, and simultaneously compete for the collective recognition of their products as legitimate and significant novelties for collective aesthetic/epistemic purposes.

In this chapter we trace the historical processes through which the arts and sciences became institutionalized as fields of novelty-driven cultural production in which the existence of collective aesthetic/epistemic purposes, the mutual awareness of each other's production, responsive practices of novelty production and the competition for collective recognition *became taken-for-granted features of artists' and scientists' creative work*, and through which the modern socio-economic contexts of their creative production became established. We trace these institutionalization processes for the arts and sciences in Europe, where they occurred for different arts and sciences at different times and with different speed in different societal contexts. Although similar variations in these institutionalization processes can be found outside Europe, it is impossible here to produce an exhaustive account. However, it is possible to demonstrate how this process took place for some of the arts and sciences in Europe between the 16th and 19th centuries, and thus to reconstruct the logic of this process and its result, at which all of the arts and sciences arrived, if at their own pace.

The institutionalization of the arts as novelty-driven fields

The institutionalization of the arts as novelty-driven fields of cultural production involved the separation of artists from craft-based artisans and of aesthetic purposes from specific social functions and occasions in what Larry Shiner has characterized as *The Invention of Art* as a distinct social system of ideals, practices and institutions in his 2001 book of that title. In particular, it established the role of the artist as the independent creator of original works in freedom from: (a) the imitation of traditional models, (b) the dictates of reason and rules, (c) restrictions on fantasy and imagination and (d) the exact imitation of nature (Shiner, 2001, p 112).

These processes have been understood by some as originating in the Italian Renaissance. For instance, Ernst Gombrich (1966) has suggested that Lorenzo Ghiberti's demonstration of the perspective construction of a round building in *The Story of Joseph* low relief panel in his second door for the Florence Baptistery between 1425 and 1452 represented an intellectual problem solution that would be admired and valued by his fellow practitioners and connoisseurs as a significant contribution to the collective body of knowledge and practices. As such, it could be seen as the embodiment of a radical change in the prevalent frame of reference within which Ghiberti was working, going beyond achieving the high standards of the past to make progress towards newer purposes and ideals in a similar way to the emerging institution of science and its belief in the infinite progress of knowledge.

However, it is not clear that this achievement did indeed constitute an exemplary innovation that heralded the separation of the intellectual pursuit of Art from the applied arts of the craftsmen rather than being an unusual and ingenious way to deal with a common problem faced by artisan/artists in representing Biblical events in low relief. Insofar as Gombrich's claim relies on Ghiberti's own account of the history of modern art from the innovations of Giotto to his autobiography in his *Commentarii*, it is worth noting that this manuscript only survives in one case and never seems to have been brought to the point where it could be considered as a coherently publishable book for dissemination to fellow practitioners as distinct from functioning more as a family memoir. As Martin Kemp has emphasized:

> Ghiberti's compilation as a whole seems not to have been primarily intended for broadcast through presentation to a patron or for circulation in the workshops. Rather it tends to assume the guise of a treasury of knowledge, containing the intellectual 'secrets' which he wished to bequeath to those worthy of receiving it. His 'secrets' do not just reside in the highly technical procedures of bronze casting, gilding and so on, which would be passed down in the workshop, but also in a profound sense of the intellectual artist in the grand span of history and of the arcane wonders of visual science. (Kemp, 1997, p 107)

Similarly, Shiner (2001, p 45) has pointed out: 'the normative practices and concepts of the artisan/artist in the Renaissance were still far from modern ideals of the (artist as the) autonomous creator', stating that:

> Renaissance painting and sculpture were not evaluated in their own time by modern standards of originality but by whether they gracefully overcame the difficulties of invention and execution … Michelangelo's practice and outlook were closer to that of his fellow artisan/artists than to that of the modern artist who places self-expression and originality above skill and service. (Shiner, 2001, p 47)

Furthermore, Michelangelo's claimed freedom to paint the ceiling of the Sistine Chapel as he wished did not extend to choosing the theme and purpose of this work on his own.

> Michelangelo did not mean that he could paint whatever he wanted, rather, like Ghiberti, he meant that he could treat his theme in whatever manner he chose. Probably too, like Bellini he meant that the theme could not be constricting, but rather must yield when necessary to pictorial invention, so that the distance between conceptual and visual elaboration was closed. (Summers, 1981, p 453)

Patronage remained critical to the development of careers as artisan/artists well into the 17th century as Haskell (1980) has documented in his analysis of patronage in Rome and Venice in that period.

Another suggestion that changes in artistic practices and purposes during the Italian Renaissance were developed in a comparable manner to those in the modern arts and sciences has been made by Ruth Katz (1984) in her analysis of the role of informal groups of musicians, amateur enthusiasts and theorists of music under aristocratic sponsorship in late 16th-century Florence. She considers these 'academies' to have functioned as collective communities dealing with specific artistic problems in a similar manner to the invisible colleges of researchers working on scientific ones in the modern sciences (Crane, 1972).

Two of these groups, the *Camerata* and *Alterati*, were important arenas for debate and collective problem solving in the composition and performance of music that contributed to the development of opera as a distinct musical genre. Members of these groups were concerned to reform the polyphonic music of the period as part of discovering 'the ideal combination of words and music such that text and music, each in its own way and in juxtaposition to the other, could be maximally effective in communicating … the specific meanings and emotions to the text' (Katz, 1984, p 67).

In dealing with these problems 'the group of composers and poets loosely allied with one another began writing and publishing substantial narrative theatrical works in which all the characters sang and sang all the time', creating a 'minor earthquake' in musical history (Abbate and Parker, 2012, pp 40–2) that contributed to the 17th-century development of opera as a business run by impresarios for a diverse paying audience, initially in Venice (Abbate and Parker, 2012, pp 48–53).

However, such groups in Renaissance Florence encompassed a much wider range of roles and interests than do the invisible colleges of the modern sciences and remained dependent on the support of nobles and rich patrons, as did most artistic practitioners. Even the most successful and internationally renowned Italian painters of the 16th century, who could move from city to city without having to pay local guild membership fees or follow their restrictions on materials and prices, often had to paint family portraits, decorate rooms and design tapestries and furniture in patrons' properties. Unlike the modern artist, most Renaissance court artists made their works on demand for a specific social function rather than creating them as personal expressions. Many contracts for paintings and sculptures stipulated not only their size, price and delivery date but also the subject matter and materials (Baxandall, 1972, pp 1–27; Shiner, 2001, pp 42–3).

As Haskell (1980, p 21) has suggested in his analysis of 17th-century patronage: 'the very nature of the contracts that have been examined shows that artists were still generally looked upon as superior craftsmen'. The importance of personal patronage to artists' careers, particularly in their early stages, is exemplified by the impact of changes in papal incumbency on the fortunes of artists from different Italian cities in Rome as Popes, their nephews and other family members dispensing patronage, gained and lost access to Papal resources (Haskell, 1980, pp 3–23).

Elsewhere in Europe, painting and sculpture struggled to gain general recognition and status as liberal arts. Guild membership and restrictions remained important to artistic practices and careers until well into the 17th century. Most Dutch painters maintained close professional and social alliances with goldsmiths, glassblowers and embroiderers, often in the same guild, as in Delft (Montias, 1982, pp 33–107), and were more concerned to distinguish themselves as the most accomplished and polished craftsmen rather than claiming superiority over them as artists (Alpers, 1983, pp 112–3).

Over the course of the 'long' 18th century (1680–1830) the dominant patronage model of art production and reception began to give way to the market model in much of Europe (Bonds, 1996, pp 24–5; Shiner, 2001, p 128; North, 2013, pp 147–8). By the end of this period, audiences for what became known as the fine arts had expanded greatly and become much more differentiated, particularly for literary works in vernacular languages, than the smaller 17th-century audiences of patrons, connoisseurs and amateur enthusiasts whose demands were specific and aesthetic tastes well known. Changing audience size, diversity and anonymity forced a recasting of the terms in which the arts were conceived, particularly the boundaries between the 'polite' arts and their audiences, on the one hand, and more popular modes of entertainment, on the other hand (Shiner, 2001, pp 94–5). Audiences became large enough to support artistic production and distribution through their combined individual choices.

During this period many of the modern fine art institutions were established in Europe, albeit to different extents in different societies and fields. These included literary criticism, history and canons, art museums, the history and criticism of painting and sculpture, the public performance of secular music in concert halls, musical criticism and, somewhat later, the formation of a canon of musical works and state conservatoires. These institutions embodied 'the new opposition between fine art and craft by providing places where poetry, painting, or instrumental music could be experienced and discussed apart from their traditional social functions' (Shiner, 2001, p 88).

In the case of literature, the first copyright law, the British Statute of Anne (1710), that vested ownership of a manuscript with its writer signalled a radical change in the status and role of authors and their ability to make a living from creating literary works. Beforehand:

> Whether as a craftsman or inspired, the writer of the Renaissance and neoclassical period is always a vehicle or instrument: regarded as a craftsman, he is a skilled manipulator of predefined strategies for achieving goals dictated by his audience; understood as inspired, he is equally the subject of independent forces, for the inspired moments of his work … are not anymore the writer's sole doing than are its more routine aspects, but are instead attributable to a higher, external agency – if not to a muse, then to divine dictation. (Woodmansee, 1994, pp 36–7)

However, over the course of the 18th century he became 'a unique individual uniquely responsible for a unique product' (Woodmansee, 1994, p 38).

As such, the writer could sell his or her work to lending libraries and other intermediaries for distribution to a growing anonymous audience and so publicly affirm both ownership and authorizing power. Together with the rapid expansion of the market for books and journals, the establishment of copyright hastened the separation of imaginative from general literature and enabled writers to make part of their living from literary criticism as consumers sought advice on what to read amid the flood of new publications (Shiner, 2001, pp 88–9). It also supported the development of a literary community of authors that debated and demonstrated new approaches to narrative fiction in the emerging genre of the novel (Watt, 1957, pp 54–9, 290–301; McKeon, 1987, pp 410–9).

In contrast, copyright laws were not passed in France and Germany until the end of the 18th century and patronage remained critical for writers to survive as producers of imaginative literature. In France, the combination of censorship and tight restrictions on the number of printers forced many writers to depend on clandestine publishers until the revolutionary regime passed the *Chenier* Act in 1793, and most men of letters relied on small inheritances, occasional patrons and intermittent income from print. As Clark (1987, p 44) suggests: 'Patronage could define literary life because no real alternative sources of support existed. Prior to the great expansion of the market in the nineteenth century, writers could not hope to realize substantial income from the sales of their books'.

Furthermore, the prevailing ideal of cultivated amateurism and integration of literary and elite activities inhibited the development of a separate professional ethos in pre-revolutionary France. In Germany, the combination of the lack of an effective copyright law before 1810, endemic pirating of books in the numerous independent states and the widespread view that writing was largely a vehicle for expressing received ideas already in the public domain meant that the rapid increase in the reading public after 1770 failed to support the institutionalization of the professional author role (Woodmansee, 1994, pp 40–54).

In addition to copyright laws, the rapid growth of printers, booksellers and other means of distributing printed materials greatly expanded the demand for new writing, particularly in England. After the Glorious Revolution in 1688, the 1695 lapse of the Licensing Act and the removal of the Stationers' Company's monopoly of control over printing and publishing, pre-publication censorship ceased and the number of printers mushroomed across the country with 400 outlets operating in 200 towns in 1740 and nearly 1,000 in over 300 locations by the 1790s. From producing 21,000 titles in the 1710s they turned out over 56,000 by the 1790s. As well as the number of booksellers catering to the new middle classes also growing, new means of accessing the flood of books and other materials were established, with about 100 circulating, proprietary and subscription book clubs and libraries operating in London by 1800 and a thousand elsewhere (Porter, 2000, pp 72–95).

Such growth encouraged the specialization of both authors and audiences, with men of letters seeking to distinguish themselves from the denizens of Grub Street catering to the taste of the mass readership and the more educated readers denying their indulgence in sentimental novels (Porter, 2000, pp 282–94). After the disappointing reception to the publication of *Lyrical Ballads* in 1798, Wordsworth concluded that great and original writers could not expect their immediate audience to understand their new works but had to work to create the taste for appreciating them (Woodmansee, 1994, pp 38–9), highlighting the creative originality of artistic genius (Shiner, 2001, pp 112–5).

The status of painters and sculptors in France was significantly altered by the 1648 establishment of the *Académie Royale de Peinture et de Sculpture*, which officially subordinated 'a still-medieval organization of artistic practice, craft-oriented and mercantile ... to one in which the most conspicuous features and concerns of guild life were not only absent but specifically forbidden' (Crow, 1985, p 25) and institutionalized a regular school of figure drawing in an effort to achieve higher artistic standards in a similar manner to the role of the Academy of St Luke in Rome.

Although the new Academy struggled to establish itself during the next two decades of political turmoil, it gained a monopoly of control over state patronage and formal artist training in the 1660s as part of Colbert's consolidation of state control over high culture. However:

> we need to be wary of substituting bureaucratic intentions for actual outcomes, and the official imposition of an ideological position cannot instantly create a genuine community imbued with these principles ... the ideal of a codified, intellectually articulate, and publicly scrutinized collective practice remained no less a minority position. (Crow, 1985, pp 28–9)

After a number of public exhibitions of paintings were organized by the Academy over the next several decades, in 1737 it began to hold regular exhibitions of its members' works in the *Salon Carré* of the Louvre that displayed contemporary art for the first time for free in Europe. These Salons stimulated the growth of a varied middle-class audience for new paintings as well as journalistic criticism and discussion about their subject matter and styles. The number of similar Academies (of the arts) grew considerably across Europe in the late 18th century, from ten in 1740 to over a hundred by 1790, usually enjoying royal protection and holding regular exhibitions of their members' works. One example was the British Royal Academy of Arts (1768–69) which was intended, among a number of purposes, to raise the status and accomplishments of artists in a comparable manner to the Roman and French Academies (Shiner, 2001, pp 101–2; Smith, 2012).

Additionally, the number of collectors of paintings in France grew considerably over the course of the 18th century, from around 150 in 1700–20 to at least 500 between 1750 and 1790, accompanied by a rapid increase in the number of annual auctions of paintings and in the status of art dealers who began to specialize in the display and sale of easel paintings rather than of furniture, jewellery and other domestic goods. Together these changes resulted in the institutionalization of major components of the modern French art market by the middle of the century (Pomian, 1987, pp 147–60). Internationally, this period also witnessed the establishment of many national art museums based on royal and aristocratic art collections from the 1733 establishment of the Capitoline Museum in Rome onwards that became open to the public and helped to disseminate the emerging canon of western European art as well as being a resource for training new practitioners (Paul, 2012).

This expansion in the size of the art market, the frequency with which paintings were bought and sold, and the public visibility and prestige of paintings and sculptures encouraged both considerable growth in the number of artists and their separation from house decorators, furniture painters and tapestry designers as specialists in the production of easel paintings for sale to a largely anonymous audience. By the 1750s sales catalogues and the Salons began to list paintings by artists' names as interest grew in individual styles and signatures (Pomian, 1987, pp 156–9).

In contrast, court, church and aristocratic patronage remained important for the composition and performance of music in most European countries throughout the 18th century, although audiences for some operas, particularly *opera buffa*, did include members of the bourgeoisie and the popular classes in Italy, and, later, elsewhere as performances became more dependent on public ticket sales (Abbate and Parker, 2012, pp 49–53, 120–5). Apart from Händel, who was able to acquire considerable wealth through his entrepreneurial activities as well as writing, conducting and performing his own works in London and elsewhere, albeit with some royal support (Scherer, 2004, p 4), most composers were members of aristocratic households or religious institutions performing a number of roles for particular occasions at the behest of their employers. Partly because of the lack of copyright protections for musical works, the costs involved in printing and distributing scores, and the limited availability of musicians and spaces for their public performances, the incentives and rewards for producing them without patronage remained rather limited for much of this period.

Composers such as Mozart and Beethoven often had to act as their own impresarios if they were to have their music performed in public (Elias, 1993; Scherer, 2004, pp 62–3, 74–5). Haydn's economic and social success in writing music for London audiences in relatively large rooms towards the end of the end of the century was notable by its atypicality and he had to return to Esterhazy when his patron demanded his presence (Goehr, 1992, p 210). Indeed, the production, ownership and reification of distinct musical works that could be reproduced, distributed and performed to multiple audiences in different locations as the same entity did not become widely institutionalized as a separate set of practices until the 19th century. Similarly, the role of composer continued to overlap with that of performer and conductor until the end of the 18th century so that they were not recognized as the creators of musical works until the following period, as discussed by Lydia Goehr in her *Imaginary Museum of Musical Works* (1992).

These changes in the institutions governing the production and dissemination of art works to different audiences in the long 18th century hastened the decline of neo-classical ideals and standards in the arts and supported the diffusion of Romantic ideas across Europe from roughly the last few decades of the 18th century to the mid-19th century. Romanticism exalted the creative imagination of individual artists in expressing intense personal emotions, often in response to experiences of natural phenomena, as well as the genius of individual scientists in understanding the natural world through careful observation and experiments aimed at uncovering hidden forces such as electricity and chemical reactions (Holmes, 2008). Rejecting many of the established tenets of Enlightenment rationalism, it emphasized the individual basis of inspiration and understanding as well as preferring organic conceptions of natural processes to mechanical ones and opposing reductionist hierarchies of the sciences. Because of the insistence on individual inspiration and experiences being the basis of authentic works of artistic genius rather than the classical goals of imitating nature and established models of beauty, originality of expression was regarded as a necessary feature of great art works. However, pursuing novelty for its own sake in isolation from previous conventions and achievements was regarded as insincere and meretricious.

Over the course of the 19th century, the audiences for art works and performances expanded greatly in many European countries and elsewhere, boosting demand for, and analysis of, new contributions. This growth was aided by a number of significant changes in the socio-political context of the arts. These included: (a) the expansion of primary and secondary education, (b) the further establishment of public art galleries and museums, including some focused on contemporary paintings and sculptures (Lorente, 2011; Paul, 2012), (c) improvements in paper making and printing technologies that significantly reduced the costs of producing papers, magazines and books (Febvre and Martin, 1976, pp 33–7), (d) reductions in taxes on publications and state censorship (Clark, 1987, pp 52–5; Cottington, 2022), and (e) the creation of public concert halls and other forms of support for musical performances (Weber, 2008), together with the considerable increase in the number of educated civil servants, professionals and business owners interested in the arts (Weber, 2003).

Audiences also became more internationalized as contemporary art works were increasingly bought and sold across national borders. In addition to spreading literacy throughout the bulk of the adult population and enabling general access to art works, these changes provided the basis for increasing the number of people with the economic and social resources for actively participating in the appreciation and understanding of art works beyond the earlier presumption that this informed audience was restricted to the aristocracy and educated members of the leisured classes.

Additionally, more states established and expanded national Academies, conservatoires and other forms of training artists as well as a variety of prizes, positions and other resources for supporting artistic careers and, eventually, the emerging field of the history of art, partly stimulated by the growth of nationalist ideologies and identities (Brush, 1996; Moxey, 2001). Together with the striking success of some practitioners in literature, painting and music in reaching large audiences, these developments encouraged an increasing number of educated people to try to establish careers in the arts (Clark, 1987, pp 57–60). This growth provided the basis for an enlarged audience of well informed and highly interested neophyte artists, critics and disseminators of art works and commentaries about them. Audiences thus became considerably more differentiated in terms of their interest in, and knowledge about, the production and distribution of art works, and aesthetic conventions. This growth of aesthetically informed audiences provided the basis for artists to develop innovative approaches and styles that challenged established standards and ideals.

The growth of the market for artistic works and experiences and associated expansion of their creators seeking recognition as artists in the 19th century intensified competition between them for access to audiences and influence over aesthetic ideals and practices. As Priscilla Clark (1987, pp 79–95) has emphasized, in the case of French literature this competition encouraged the development of collective strategies to gain attention and authority by increasingly cohesive groups of writers, from the loosely organized romantics in the 1820s and 1830s to the more closely knit Parnassian poets in the 1860s and 1870s and the tightly integrated naturalists under Zola's leadership in the 1870s and 1880s. These groups were often allied with particular publishers and critics supporting their distinctive aesthetic principles and ideological identities in opposition to established conventions. However, at the same time as drawing attention to their innovative approaches to literature of offering something new to the market, partly by producing novels as well as poems and plays that were traditionally more prestigious genres, French writers continued to seek admission to the *Academie Francaise* and elite Salons and assert their fidelity to long standing artistic ideals.

Similar collectivities of artists seeking to assert and promote distinctive aesthetic ideals and practices were organized in other genres, such as the Brotherhoods of painters discussed by David Cottington (2022, pp 21–45). While some of these were largely conservative, elitist and esoteric groups opposed to both the burgeoning commercialization of artistic production and distribution and the growing authority of national Academies in the early 19th century, later on they engaged more actively with market institutions and created collective identities that helped to gain the attention and support of the expanding middle classes, such as the Pre-Raphaelite Brotherhood in England.

Competition between artists for general audiences' attention and support was increasingly extended to competition between interpreters and disseminators allied to distinct aesthetic ideals and standards for the support of more knowledgeable and interested audiences. Criticism of new art works became increasingly differentiated between general reviews of new novels, exhibitions and other outputs, often written by journalists, and more specialized and technical analyses by professional critics who focused on particular genres and their histories and took part in current debates about aesthetic and practical problems for artists and their more aesthetically informed and interested audiences (Prettejohn, 1997).

A less aesthetically directed collective movement of artists developed in many European countries in the 1880s and 1890s following the declining prestige and authority of state Academies as the dominant organizations controlling the training and certification of artists and the display of their works in nationally organized Salons. This 'Secessionist' movement has become seen more as the cumulative result of a variety of attempts to gain more practitioner authority over the public display and recognition of new works as part of artists' wish to establish a professional identity and socio-economic status for fine artists rather than as an avant-garde promoting radically novel aesthetic ideals and practices in systematic opposition to the authority and procedures of established institutions. Indeed, many participants in these movements sought public support for their independent exhibitions and adopted a more *juste milieu* or middle of the road aesthetic position rather than promoting a complete break from official ideals and practices (Jensen, 1994, pp 167–200; Cottington, 2022, pp 86–106).

The development of the Secessionist network of artists and dealers across Europe facilitated the emergence of the avant-garde movement dedicated to radical aesthetic goals and practices by further weakening the dominant institutions governing the training and certification of artists and their access to audiences. Particularly in France, the combination of the expansion of private 'academies' for training neophyte painters and sculptors, the proliferation of *petites revues* dedicated to the analysis and promotion of literary and artistic innovations, and the emergence of dealer-centric and profitable markets for innovative art styles in Europe and North America, helped to develop the infrastructure for greater artistic autonomy and what Cottington terms the alternative professionalism of the avant-garde (Cottington, 2022, pp 130–47). Unlike the artists pursuing more middle of the road aesthetic strategies, these '*novateurs*' prioritized commitment to specific and innovative aesthetic principles and practices, such as the innovations adopted by many Neo- and Post-Impressionist painters, and promoted a succession of 'isms' in painting (Cottington, 2022, pp 133–41; see also the case study of Cubism in Chapter 5).

These commitments to developing new ways of conveying meaning and representing experiences encouraged an increasing concern with self-consciousness and self-referentiality in constructing art works. Creative practitioners became more concerned with exploring the formal properties of each medium, its purity as a means of expression such as abstraction in painting, atonality in music and experimentalism in literature, liberating art from social functions and established conventions. The Russian Formalists, for instance, argued that what makes a poem or novel literary are its self-referential features of pattern, image, trope and rhythm rather than the expression of emotion or dramatic narratives (Ehrlich, 1973; Everdell, 1997, pp 347–60; Shiner, 2001, pp 247–8; North, 2013, pp 145–69). This concentration on the means of expression, rejection of established conventions and the concomitant need to develop novel forms further reinforced the importance of originality in the arts, especially developing new ways of creating art works. By focusing on the distinctive qualities of each medium and how artists could interpret these anew in expressing their ideas, artists became more attentive to each other's formal innovations and more collectively organized around competing aesthetic norms and practices.

The institutionalization of the sciences as novelty-driven fields

The institutionalization of the sciences as novelty-driven fields of cultural production took place over roughly the same period as that of the arts. It began in Europe at the end of the 15th century and was only completed for certain fields as late as the 19th century. However, while the practices for producing and assessing works in different arts/crafts were already established, the epistemic problems addressed by the sciences and the practices through which these problems were addressed changed fundamentally with their transformation into novelty-driven fields. This occurred through: (a) communities beginning to produce knowledge collectively by using and building on each other's contributions, (b) a consolidation of practitioner communities around specific sets of epistemic problems, which were often linked to circumscribed domains of empirical investigation, and specific practices of addressing these problems, and (c) the transformation of relations to society through increasing professionalization of scientific work and dependence on the state as the main funder of knowledge production.

This complex process can be traced by dividing the period of transformation into two overlapping phases and analytically separating major developments in each one. In a first phase, which began in the 15th and lasted at least until the mid-18th century, several processes that contributed to the institutionalization of the sciences were set in motion. These included a major transformation of the domains of knowledge production, with the subject matter of knowledge production changing and providing grounds for the later emergence of scientific disciplines. At the same time and often concomitant with the reorganization of knowledge domains, new ways of producing knowledge were introduced, even though they did not yet become dominant and universally accepted.

Changing domains and practices were accompanied by two further processes that contributed to the social reorganization of knowledge production: the emergence of new roles in knowledge production – generalized behavioural expectations tied to social positions in the knowledge production process – and the emergence of new identities – self-images of producer's positions and practices. Neither of these processes was completed in the first phase, and they began and evolved at different times and at different rates of development in particular domains of knowledge production and geographical regions.

In a second phase, which began for some fields of knowledge production in the late 17th century and lasted for some through the 19th century, knowledge production became consolidated in specialized communities addressing shared epistemic problems related to distinct bodies of knowledge. This process completed the institutionalization of novelty-driven production because in these specialized communities – scientific disciplines in the modern sense – contributions were produced with increasingly standardized practices, and standards of relevance and quality began to emerge. By addressing shared epistemic problems and applying shared standards, the relevance of others' contributions increased despite them also being competitors because without responding to them in some way, producing relevant contributions became impossible.

Both periods have been termed 'scientific revolutions'. The transformation of the sciences in the early modern period, from ca. 1490 to ca. 1730, has for a long time been discussed by historians of science as 'The Scientific Revolution'. From the 1990s onwards, this understanding has been challenged on the grounds of the change in that period being gradual rather than abrupt and too complex to be reduced to one single revolutionary event (Shapin, 1996; Osler, 2000; Park and Daston, 2006). Other historians consider the concept of a 'scientific revolution' still useful and use it without reservations (for example, Porter, 1996; Dear, 2006). The application of the term 'second scientific revolution' to the second phase by several authors (Kuhn, 1977; Cohen, 1985, pp 91–101; Golinski, 1998) has encountered less resistance, probably because the term has never become as imposing as 'The Scientific Revolution'. Regardless of these terms, historians of science agree on 'huge, and hugely varied' transformations in the first phase (Park and Daston, 2006, p 13) and on the emergence of modern scientific disciplines in the second phase (Porter, 2003).

From the perspective developed in this book, applying the term 'scientific revolution' to both phases is justified because the institutionalization of novelty-driven production included a fundamental transformation of the content and of the social organization of knowledge production. Prior to the changes traced here, producers of knowledge[3] already worked relatively autonomously but often developed problems and solutions separately from each other. While contributions were framed by referring to classics and contemporaries, the main purpose of these references were legitimization (by reference to the classics) and competition (with contemporaries). How an author related their work to that of others was a matter of individual choice rather than by following collective standards, and the further development of others' contributions was not the dominant form of knowledge production. With few exceptions that will be discussed later, members of these communities perceived themselves as engaged in similar practices but not as jointly producing knowledge. They corresponded extensively and visited each other but could make legitimate contributions without directly responding to the work of others. Interactions were often competitive and mainly concerned the question of who had the better arguments.

The two scientific revolutions transformed this situation by institutionalizing the production of scientific knowledge as an enterprise in which successful participation depends on constructively responding to others' production of new knowledge. This institutionalization process took several centuries, and its initiation and speed varied greatly across fields of knowledge production and geographical regions despite the knowledge production being already internationalized to some extent. The account of institutionalization processes is therefore a reconstruction of main developments which is illustrated by examples rather than a generalized history of the development of the sciences in the period under discussion.

The transformation of knowledge domains

In the 16th and 17th centuries, the domains about which knowledge was produced became extended and restructured. Some domains of knowledge production became transformed by a fundamental cultural shift that: (a) considerably enlarged the range of phenomena that were considered natural, (b) led to a general understanding of natural phenomena as having natural causes and (c) considered phenomena and their causes to be the same everywhere. According to Daston (1998, pp 149–56), in the beginning of the 16th-century conceptions of phenomena that were not considered natural included the supernatural (suspensions of the natural order by signs and miracles, which were created by God), the preternatural (phenomena 'beyond nature' created by a variety of agents including God, angels, demons and others), the artificial (things made by humans) and the unnatural (a moral category that designated heinous crimes like patricide and bestiality).

The number of phenomena considered supernatural was reduced when miracles came under intense scrutiny in the context of the struggle between Catholics and Protestants and because the direct invocation of God through miracles was used as, or seen as, undermining the authority of the state and/or established religions. The preternatural, which included monsters (abnormal births), strange weather, figured stones, petrifying springs, the occult virtues of plants and minerals, and myriad other deviations from the ordinary course of nature (Daston, 1998a, p 155), was highly popular in the mid-16th century but became subsequently 'naturalised' for the same reasons (Daston, 1998a, p 163). For example, 'monsters' (abnormal births) were considered divine prodigies at the beginning of the 16th century and 'had been integrated into the medical disciplines of comparative anatomy and embryology' by the end of the 17th century (Park and Daston, 1981, p 23).

This cultural shift made natural phenomena accessible to investigation and thus extended the domain about which knowledge could be produced, and phenomena that required explanation. It largely eliminated God, angels, demons and magicians as causes of natural phenomena. Phenomena previously considered as 'preternatural' became 'naturalised', and work in jurisprudence and theology introduced the idea of 'laws of nature' that had been imposed by God and could be investigated (Park and Daston, 1981; Dear, 1990; Daston, 1998a).

The extension of the empirical domain of inquiry and explanation coincided with the integration of four previously separated bodies of knowledge and practices of knowledge production. Of these, *natural philosophy* was understood as the science of natural bodies and sought the universal causes of phenomena related to them (Garber, 2006, p 26; Blair, 2006). Together with other pursuits, natural philosophy belonged to *Scientia*, which 'referred to any rigorous and certain body of knowledge that could be organized … in the form of syllogistic demonstrations from self-evident premises' (Park and Daston, 2006, p 4). These premises or axioms 'had to be rooted in familiar and commonly accepted *experience*' (Dear, 2006, p 109, italics in original; see also Dear, 1987, p 141; Park and Daston, 2006, p 3). An example of such a premise would be 'the sun rises in the east', which 'was unshakably and universally known to be true through experience' (Dear, 2006, p 109). This approach applied to natural philosophy because it was part of *Scientia*.

Distinct from natural philosophy was *natural history*, which was rooted in an ancient form of knowledge production that is 'most closely associated with the writings of Pliny the Elder (ca. 22–78)' (Findlen, 2006b, p 437). Pliny's writings represent an attempt to describe all entities found in nature, or derived from nature, that were accessible in the Roman world. His contemporary Dioscorides (ca. 40–80) focused on the medicinal efficacy of plants and implicitly suggested that 'any descriptions of nature should always be in the service of medicine' (Findlen, 2006b, p 438). Only in the Renaissance did the description of nature become a distinct and separate domain of knowledge production (Ogilvie, 2006). Building on the classic tradition which reached back to Pliny's predecessors Aristotle and Theophrastus, natural historians collected and attempted to systematize knowledge about natural objects like animals, plants and minerals. In contrast to natural philosophy, natural history was not concerned with explanations of natural phenomena (Findlen, 2006b; Ogilvie, 2006; Park and Daston, 2006, p 4).

The *mathematical sciences* formed yet another distinct and separate body of knowledge and practice. They did not search for causes of phenomena either but sought quantitative descriptions of them. They were generally subdivided into pure mathematics (mathematics and geometry as dealing exclusively with quantity) and mixed mathematical fields that also dealt with quality. The latter included astronomy and astrology, music and optics (Donahue, 2006; Mancosu, 2006; Park and Daston, 2006, p 4).

Finally, natural philosophy, natural history and the mathematical sciences were distinct from the '*mechanical arts*', which 'included practical applications of mathematical knowledge in fields such as architecture, navigation, clockmaking, and engineering' (Park and Daston, 2006, p 4; Bennett, 2006). The instruments offered solutions to practical problems and were not considered as means for making discoveries. They 'were adjusted to different contexts; the astronomer's astrolabe, for example, was simplified and adapted to both navigation and surveying, as was the astronomical cross-staff' (Bennett, 2006, p 684).

These four domains 'were conceived as separate pursuits, with their own methods, goals, and intellectual as well as social status', which were by and large conducted by different people (Park and Daston, 2006, p 4). From the end of the 15th century onwards, however, their separation was eroded by several processes in which domains of knowledge production were extended, changed their purposes and merged. *Natural philosophy* as the only discipline that aimed at explaining natural phenomena became transformed by moving from Aristotelian explanations[4] to a mechanical natural philosophy that tried to explain the behaviour of material bodies from the behaviour of their fundamental constituent parts and to identify 'laws of nature' governing this behaviour. For this purpose, it developed an experimental tradition (Bennett, 1986). Experiments had been conducted in some domains of knowledge production other than natural philosophy since Antiquity, but 17th century natural philosophy in England saw a strong rhetorical promotion of experiments (Dear, 2005), with the birth of the concept 'Experimental Philosophy' (Feingold, 2016), the appointment of a curator of experiments (Robert Hooke) by the Royal Society in 1662 (Shapin, 1988, p 382) and the distinction between speculative and experimental natural philosophy becoming 'the primary methodological framework within which natural philosophy was interpreted and practiced in the late seventeenth century' (Anstey, 2005, p 216). Metaphysical discussions about the nature of causation and metaphysical foundations of physics began to become separated from physics, and the interest of natural philosophers focused on the latter rather than the former (Bennett, 1986; Blair, 2006; Garber, 2006; Joy, 2006).

Based on its collection, description and classification of specimens which extended classical empirical knowledge, *natural history* evolved from discussing the classics to checking the correctness of the translations of their works and, later, the content of their statements about nature. This interest in checking and correcting classic works triggered increasingly systematic empirical observations from the beginning of the 16th century (Findlen, 2006b). Empirical investigations were also facilitated by the institutionalization of medicinal botany as part of medical education in the 16th century and the associated growth in number of botanical gardens and of botanical field trips of students and groups of naturalists (Findlen, 2006b). The discovery of the Americas and long-distance travel in general: (a) caused an explosive growth of knowledge about remote places, (b) undermined the authority of the classics by showing some of their statements to be wrong and (c) contributed to an understanding of facts as reported (rather than experienced) true events, which in turn (d) helped establish the idea of knowledge as provisional (George, 1980; Hooykaas, 1987; Cohen, 1994, pp 151–98; Daston, 1998a). Fossil records additionally challenged the established understanding of the history of nature (Findlen, 2006b). Overall, the number and variety of phenomena that required explanation, and which were not known to the classics and thus not explained, or wrongly explained, by them, grew. This growing demand for explanations moved natural history closer to natural philosophy (Daston, 1998a; Findlen, 2006b, p 463).

The *mathematical sciences* also began to merge with natural philosophy. Astronomy became synthesized with natural philosophy through its transformation from a science of mathematical description to a science concerned with explanations of the movement of astronomical objects (Donahue, 2006). Acoustics emerged from music theory (an abstract mathematical discipline that quantitatively described music) and evolved as an experimentally oriented science of sound for which mathematics was a tool to describe the observed physical phenomena and relationships (Mancosu, 2006). Optics also was transformed into an experimental science based on experiments and searching for explanations of the observed phenomena. It became parts of the physical sciences and thus of natural philosophy (Mancosu, 2006).

Finally, the *mechanical arts* became linked to natural philosophy because the mechanical philosophy that replaced traditional Aristotelian natural philosophy 'took the machine as the paradigm of action throughout the natural world' (Bennett, 2006, p 694). They also became increasingly mathematized (Bennett, 2006). Applied mathematics thus became a common approach that linked previously separated activities of knowledge production and became an important tool for explaining natural phenomena.

New practices of creating and communicating knowledge

The transformation of the four domains of knowledge production, natural philosophy, natural history, mathematical sciences and mechanical arts, was in part driven by the increasing role of new practices of creating and communicating empirical knowledge. All four domains increasingly relied on observations as sources of knowledge about nature. The invention of the microscope and the telescope provided new opportunities to produce knowledge and undermined the traditional Aristotelian distinction between the natural and the artificial world (Van Helden, 1983, pp 50–3; Dear, 1987). Also, experiments were increasingly seen as an important means of producing new knowledge even though the ideology of experiment grew much faster than the actual practice of experimenting. Experiments had been used to produce new knowledge for a long time, for example by alchemists (Newman, 2006), and had been reported in the 'books of secrets' of the 16th century, which contained collection of recipes from alchemy and the crafts (for example metallurgy instrument making). Eamon (1984) describes the contribution of this tradition to the rise of experimental practice:

> The earliest of these works were literally 'how-to' handbooks designed by printers to provide specialists and the general reading public with useful technical information. In the second half of the sixteenth century, however, some of the virtuosi of Europe seized the opportunity presented by the expanding market for such books by collecting and publishing their own recipes and 'secrets'. By the end of the century, the so-called 'professors of secrets' had transformed the technical recipe book into a collection of scientific experiments; some even recast books of secrets into general scientific and technical encyclopedias. (Eamon, 1984, pp 113–4; see also Long, 1997; Smith, 2009, pp 361–4)

These books not only provided the grounds on which some new fields of science developed like chemistry (from alchemy), magnetism, electricity and metallurgy, they also strengthened the experimental tradition and provided a communicative link between the crafts and other producers of knowledge.[5]

Experiments were conducted in laboratories, which had been the working spaces of alchemists for a long time and grew in popularity if not in importance in the scientific revolution as controlled spaces in which sensory experiences of facts of nature were produced (Smith, 2006) even though the English term 'laboratory' began to be used for these spaces only from the mid-17th century, and was not yet routinely used at the beginning of the 18th century (Shapin, 1988, p 377). While only a few universities institutionalized laboratories (for example, the University of Leiden, Smith, 2006, pp 304–5 and the University of Oxford, Shapin, 1988, p 377), laboratories were founded by many nobles (particularly in the German territories), some of whom even worked in the laboratories themselves (Smith, 2006, p 300). The same applies to England, where 'by far the most significant venues were the private residences of gentlemen or, at any rate, sites where places of scientific work were coextensive with places of residence, whether owned or rented' (Shapin, 1988, p 378). The nature of the experiments in these laboratories changed from the application of recipes to scientific experiments through which 'matters of fact' were resolved and judged (Smith, 2006, p 303).

The beginning integration of natural history, the mathematical sciences and the mechanical arts with natural philosophy confronted the latter with a problem some of the mathematical sciences had been wrestling with for a long time: how to use local and temporary empirical experiences – 'empirical novelty' (Dear, 1995, p 7) – as resources for the advancement of science. Empirical novelties did not meet the Aristotelian standards of experience, especially not those of Aristotelian natural philosophy, which was about 'that which is always or that which is for the most part' (*Metaphysics* 1027a20–27). The 'emergence of the discrete experience as the primary empirical component of natural philosophy' that accompanied the merging of the four domains of knowledge production constituted a fundamental change in the function and nature of experience for natural philosophy, which occurred in the scientific revolution (Dear, 1987, p 134). 'Singular experiences, experimental events, were not public because they were known directly only to those few who had actually witnessed them; such experiences were in consequence questionable elements of scientific discussion' (Dear, 2006, p 121; see also Dear, 1995). For these experiences to become relevant to others, strategies to 'universalize' individual experiences were extended or newly developed. These communication strategies enabled the individual to contribute their local experiences to a shared stock of knowledge (Shapin, 1984; Dear, 2006).

The reporting of observational and experimental findings gained legitimacy from the institutionalization of the notion of 'fact' by English law and the accompanying cultural change in 16th- and 17th-century English society. An event was considered as a fact if a report of it was trustworthy because: (a) the person who reported it was trustworthy, (b) observed the event themselves and (c) could name witnesses to the account (Shapiro, 2000, 2002). This understanding was also applied in England and beyond to facts that were not accessible to everyone but resulted from experiments, instrument-based observations or geographical discoveries. Some protagonists of the new forms of knowledge production were lawyers and acted as conduits for the new understanding of facts (for example, Francis Bacon and Fabio Colonna, Findlen, 2006b, p 461).

In addition to traditional practices of some of the mathematical sciences for establishing trustworthiness, such as the presentation of astronomical observational data as 'a means of generating, via geometrical models of celestial motions, predictive tables of planetary, solar, or lunar positions' (Dear, 2006, pp 121–2) and the emphasis on the repetition and repeatability of experiences (Dear, 1991; Dear, 2006, pp 122–3), new practices of establishing facts became part of the repertoire of the new communication strategies. Experiments were conducted before witnesses who confirmed them with their signature and were mentioned in experimental reports (Shapin, 1984, pp 487–9; Bazerman, 1988, pp 73–5). Observational reports listed other persons who witnessed the observed phenomenon or could serve as witness for other reasons (Fontes da Costa, 2002, pp 276–83).

Reports of observations and experiments developed as new publication genres. The scholars at the French *Académie des Sciences* created a new type of written communication – the research report – to communicate their findings (Holmes, 1991). According to Holmes, 'papers read by members of the Academy to their fellow Academicians at their weekly meetings and published subsequently as the *Mémoires* of the Academy became, by the early eighteenth century, the most consistent exemplars of the form that has persisted as the modern research paper' (Holmes, 1991, pp 166–7).

One of the reasons for this format to be developed at the French Academy was that the Academy organized sustained experimental investigation. Similar developments occurred at the Royal Society in London, where experiments were conducted and discussed regularly (Shapin, 1988; Hall, 1991).

For the contributions of practitioners to be able to affect the work of others, they needed to be accessible to all of them. This required both a change of the attitude to knowledge and changes in the practices of communication, both of which began in the 16th century. A first important change occurred in attitudes to the secrecy of knowledge. Prior to the first scientific revolution, much knowledge was not publicly accessible for a variety of reasons.

Natural philosophers and alchemists often communicated in esoteric languages in order to keep the knowledge safe from the uninitiated (Principe, 2013). The knowledge that was orally communicated in teacher–pupil relationships of alchemists and craftspeople was not accessible to third parties. Humanists, craft elites and renaissance scholars communicated through personal letters (which were not accessible to third parties) and books (which were accessible at universities but expensive to buy, both as manuscripts and later as the first printed books). Guilds and Renaissance engineers did not communicate their knowledge for economic reasons, namely, to prevent plagiarism and competition that would endanger the value of their services. Engineers who were careless in this respect and published their inventions did indeed experience plagiarism (Eamon, 1985, pp 327–8). Craft elites and renaissance scholars publishing books might also have delayed publication in order to perfect inventions (Eamon, 1984, 1985, 1991; Long, 1991a, 1991b, 2001; Iliffe, 1992, p 37; Büttner et al, 2002).

The disadvantages and constraints of open communication were eroded through several concomitant processes. Book printing offered opportunities to distribute knowledge to large audiences, which in some cases met rapidly increasing demand for knowledge (as was the case for mining around 1500, Long, 1991b, p 325). The gradual disappearance of guilds and their institutions of secrecy provided opportunities to meet this demand by publishing guild secrets. The growth of book printing also created a financial interest of printers in acquiring publishable knowledge, which made them recombine and republish previously printed content. This contributed to the literary tradition of 'books of secrets' described previously. The institutionalization of intellectual property rights as limited monopolies for the exploitation of discoveries was a response to the observation that knowledge could be put to practical use and that secrecy retarded both the advancement of knowledge and possible profits to be made from the application of that knowledge. Printers and authors began to demand the same rights, which led to the institutionalization of patents (with the earliest known patent issued in 1421, Eamon, 1985, p 328) and privileges for printed books (temporary exclusive rights to print a book, realized through the prohibition of a book being printed by others and being imported for a certain period of time) in the 15th and 16th centuries (Eamon, 1985, pp 327–9; Long, 1991a). These privileges for authors and printers removed major reasons for secrecy.

These changes increased the willingness of practitioners to share their knowledge and the ease with which knowledge became communicable, although openness remained a complex negotiated issue (Long, 2001, p 248). The necessary infrastructure for such communication improved considerably in the 16th and 17th centuries due to innovations in postal services, overseas travel and printing. The overlapping correspondence networks of scientific practitioners were made possible by the evolving regular and reliable networks of correspondence between diplomats, within religious orders and international trading houses, or through independent postal services in the 16th and 17th centuries. The intensifying of communication between European scholars gave rise to the ideology of a 'Republic of Letters' in which ideas are freely exchanged between scientific practitioners regardless of national, religious and personal differences (Daston, 1991; Ogilvie, 2016). Information from correspondence networks became printed, a link that prepared the way for journal publications (Harris, 2006, pp 347–55). The first two scientific journals were founded in the second half of the 17th century, although journals only became a major means of communicating contributions to knowledge at the end of the 18th and the beginning of the 19th centuries as specialized journals of evolving disciplines (described later). Additionally, printed books made it possible to assemble libraries which provided access to an identical foundation of knowledge in many locations.

The idea of scientific progress and collective identity of knowledge producers

The two major changes occurring in the scientific revolution of the 16th and 17th centuries discussed so far included a transformation of the domains of knowledge production due to: (a) an extended understanding of nature as an object of inquiry that is the same everywhere, (b) the consideration of all natural phenomena as requiring explanation and (c) the integration of formerly separated bodies of knowledge about nature, and the institutionalization of practices of knowledge production, which (d) contributed individual empirical experiences like findings of observations and experiments as new knowledge and (e) enabled and improved the communication of these contributions. These changes also altered the general understanding of the purposes and practices of knowledge production and the roles and identities of producers of knowledge. A core element of the new collective identity of producers was the idea of scientific progress, which included a change in both the understanding of purposes of knowledge-producing activities (to collectively advance knowledge) and the understanding of the roles of practitioners (to make contributions that are relevant to others' activities in advancing knowledge).

The origin of the idea of scientific progress was first discussed by Zilsel (1945) in the context of his Marxist argument for the importance of the educated crafts in the scientific revolution (Zilsel, 1941, 1942). He argued that the idea of scientific progress emerged in the elite crafts, that is in those crafts whose trade required education, and that this idea was transferred to natural philosophers through meetings in their workshops and sometimes cooperation. After an initial backlash (Hall, 1959, 1963), the historiography of the scientific revolution has accepted the gist of Zilsel's argument, particularly that the educated crafts played an important role in the changes of knowledge production in the 16th and 17th centuries (Rossi, 1970; Long, 2001; Bennett, 2006; Smith, 2009, p 362). Zilsel's (1945) idea that the publication of former guild secrets and of improvements of solutions to practical problems led the elite crafts to characterize their publications as contributions to a collaborative enterprise of advancing knowledge has meanwhile been corroborated. For example, Bennett (1986) has described the important role of instrument makers for both natural philosophy and the mathematical sciences and Long (2001, p 243) found that in the 16th century, '[a]uthors from diverse walks of life emphasized the value of openness for the progress of knowledge'.

While Zilsel's contention that an idea of progress of knowledge through the cooperation of individual producers of knowledge emerged in the elite crafts seems to have gained substantial support, more recent analyses suggest similar developments in most domains of knowledge production. Progress in knowledge production became recognized in *natural history*, which witnessed a rapid growth of empirical knowledge about nature after turning from the correction of classical records to their extension (Findlen, 2006b). In the *mixed mathematical sciences*, improved experiments and new observations with microscopes and telescopes not only enabled empirical knowledge to grow but also increased demand for better explanations of natural phenomena (Blair, 2006, pp 388–9). The *mechanical arts* were constantly concerned with: (a) the improvement of instruments, (b) growing collaboration between designers of scientific experiments, (c) craftspeople improving their methods, (d) naturalists attempting to relate their observations better and (e) *natural philosophers* extending their knowledge through experiments, thus making producers of knowledge aware of the extension and improvement of knowledge which they achieved through their contributions.[6]

The idea of scientific progress thus seems to have emerged in all traditions of knowledge production, be it through the recognition of the growth of knowledge or through a transition from interpreting the classics to correcting and surpassing them. The resulting collective identity of knowledge producers is reflected in the following statement by Francis Bacon:

> what may be expected from … cooperative labours and from the passage of time; especially on a road which may be travelled not only by individuals (as is the case in the way of reason), but where men's labours and efforts (particularly in the acquisition of experience) may be distributed in the most suitable way and then reunited. For men will begin to know their own strength when we no longer have countless men all doing the same thing, but each man making a different contribution. (*The New Organon*, Book I, Aphorism 113, Bacon, 1620 [2000], p 88)

This description, which was ideological and programmatic rather than a reflection of the way in which knowledge was produced at the time, was mirrored by the self-description of the Royal Society, which explicitly drew on Bacon's ideal of a cooperative investigative institution and developed a 'research ideal' of the society as 'an association of individuals with common interests who were engaged in work that was commonly perceived to be of value' (Dear, 1985, p 147). The authority of the ancient classics began to be replaced by 'the authority of the individual reporter as the actor in a well-defined, particular experience' (Dear, 1985, p 157; see also Cohen, 1994, pp 157–60; Smith, 2006, pp 295–300).

The emergence of specialized practitioner communities in the 18th and 19th centuries

At the end of the 17th century, the new ideology of scientific progress driven by dispersed specialized knowledge producers responding to each other's production was still rarely matched by actual practice. Throughout the 17th and well into the 18th century, natural philosophy continued to subordinate experiments to deduction, and new observations or experimental results were rarely theoretically embedded (Shapin and Schaffer, 1985; Chalmers, 1993; Shapin, 1994; Klein, 1996; Daston, 1998b). Reports of experiments were still rare. According to Bazerman, most communications in the *Philosophical Transactions* between 1665 and 1800 reported observations:[7]

> The most articles and pages were devoted to observations and reports of natural events, ranging from remarkable fetuses and earthquakes, through astronomical sightings, anatomical dissections, and microscopical observations. Human accomplishments received attention with accounts of technological and medical advances, travelogues of journeys to China and Japan, and an interview with the prodigy Mozart. The reportable business of natural philosophers was hardly restricted to experimenting or even theorizing, which received even less space than experiments. (Bazerman, 1988, p 65)

The discrepancy between the new collective identity of knowledge producers and the emergence of new practices and roles of knowledge production, one the one hand, and the inertia of actual approaches to the production of knowledge, on the other hand, indicates that the changes achieved in the first scientific revolution, while necessary for the institutionalization of novelty-driven production, were not sufficient. The institutionalization of novelty-driven production as the default activity of scientific practitioners required their activities to converge in specialized practitioner communities that shared a focus on sets of specific interrelated epistemic problems and approaches to solving them. Only then could the response to others' novelties become a prerequisite for producing legitimate novelties. The institutionalization of this expectation in scientific communities concerned with shared epistemic problems completed the institutionalization of novelty-driven scientific production. The main difference to the previous practice was that practitioner communities assumed collective authority over their members, who, while still autonomous in their choice of problems and approaches, must take notice of other community members' production if they wanted to make a legitimate contribution.

This second phase of the institutionalization of novelty-driven production has been analysed by historians as the institutionalization of modern scientific disciplines, and is sometimes termed the second scientific revolution. In this second phase, which began in the 18th century and in most cases extended into the 19th century, shared epistemic problems were established, and the collection or production of facts became linked to the production of specific explanations. Only in this process did the new practices of producing contributions to knowledge and of communicating them become fully institutionalized, albeit with different speed in different fields of inquiry and regions of Europe. The increasing mutual relevance of novelties for practitioners' production also led to the emergence of field-specific professional standards (Kuhn, 1961, p 190; Hahn, 1971, pp 275–6), which further strengthened the authority of practitioner communities over the production and valuation of scientific contributions. The following sketches of the consolidation of two new disciplines illustrate this process.

One of the first fields to institutionalize this knowledge-driven production and to become consolidated as a modern scientific discipline was *chemistry*, which established shared epistemic problems and approaches to solving them in the 17th and 18th centuries. Scholars adopted experimental practices and instruments from three well-established practical traditions, namely alchemy, the preparation of medicines, and metallurgy. These material cultures mainly served practical purposes in their original context but were adopted and transformed into a culture that served work on epistemic problems, 'namely, the investigation of the hidden and unobservable chemical transformations that underlay chemical operations that could be reversed' (Klein, 1996, p 255). This transformation began in the late 17th century at the French Academy, where salaried scholars developed and conducted experiments (Hahn, 1971; Holmes, 1989, 1996).

Scholars undertook numerous attempts to integrate the knowledge produced at the French academy and elsewhere in tables of relationships between elements (Holmes, 1996; Klein, 1996; Golinski, 2003, p 384). These attempts together with the isolation of what later became known as oxygen by both Scheele and Priestley independently of each other within a few years (Golinski, 2003, p 391) and the development of genuinely chemical theories such as Stahl's Phlogiston theory for the explanation of chemical processes (Klein, 1996; Golinski, 2003, pp 387–8) indicate that practitioners of chemistry worked on shared epistemic problems and responded to each other's production of novelties. The process of institutionalizing chemistry as a novelty-driven discipline can be considered as being completed with: (a) the adoption by chemists of Lavoisier's theory of combustion, calcination and acidification, (b) the even more widespread adoption of precision measurement and (c) the adoption of a new standardized chemical nomenclature, which Lavoisier developed together with Fourcroy, Berthollet and Guyton, at the end of the 18th century (Golinski, 2003, pp 393–4).

The adoption of Lavoisier' theory further separated chemical knowledge production from adjoining fields like Geology, which had used Phlogiston theory (Golinski, 2003, p 396). It thus contributed to creating the intellectual boundaries of modern chemistry. The consolidation of chemistry as a modern discipline was also indicated by the chemical textbook tradition, which can be traced back to the end of the 16th century (Golinski, 2003), specialized chemistry journals being founded from the late 18th century onwards (Nye, 1993, pp 43–6) and the foundation of professional organizations of chemists in the mid-19th century in Britain, France, Germany and the USA (Nye, 1993, pp 42–3).

Novelty-driven production was institutionalized in *physics* through its development as an experimental science which advances knowledge about specific subjects rather than all of nature. Textbooks on physics published since the early 18th century began to omit botany, zoology anatomy and physiology, which had traditionally been part of physics (Home, 2003, p 358). Experiments were conducted publicly and in dedicated university courses to illustrate scientific principles, but in the 18th century they gradually became means of producing new knowledge.

The first chair of experimental physics was created in Paris in 1753 and became a model for similar chairs throughout Europe (Home, 2003, p 358). Specialized sets of epistemic problems concerning mechanics (focusing on the mathematical description and explanation of motion), optics, magnetism, heat and electricity emerged and were advanced by new experiments. The controversies in these specialized areas (for example, about Newton's optical experiments in the 1670s, Bazerman, 1988; Schaffer, 1989) epitomize the novelty-driven nature of this work. The institutionalization of physics as novelty-driven discipline can be considered completed with the quantification and mathematization of experimental practice, which was accomplished in the beginning of the 19th century (Home, 2003, pp 371–4). The first specialized journals for the new discipline of physics were founded in 1799.

Changing social contexts of novelty-driven production

By the end of the 19th century, the domains of knowledge production that existed at the end of the first scientific revolution had all been reorganized as specialized practitioner communities in which novelty-driven production was institutionalized, and new scientific fields emerged featuring novelty-driven production. In the 18th and 19th centuries, the social contexts in which practitioner communities developed also changed considerably. This change began with the rise of a new type of scientific society in the 17th century, of which the Royal Society in England and the *Academie Francaise* are early paradigmatic examples. Against a cultural background of learned societies as the dominant form of organization of cultural pursuits, these new learned societies received royal charters, supported state governance with technical expertise and the censorship of scientific contributions through control over the publication of scientific texts by printing, and formed an institutional basis for professional careers in science. They also formalized their institutional contacts and developed a Europe-wide system of institutions (Hahn, 1975; Biagioli, 2002; McClellan III, 2003).

These new societies, whose membership overwhelmingly consisted of interested amateurs, facilitated knowledge production by providing members with an audience for the presentation of their research, publishing proceedings which became a main form of scientific communication and directing research by funding prize contests. They also undertook research projects including collaborative projects like the coordinated efforts to observe the transits of Venus in 1761 and 1769 (McClellan III, 2003, p 93).

European universities underwent major changes of their structures and roles in the production of scientific knowledge, which began in the 18th century and continued throughout the 19th century.[8] The German model of a university whose function 'was not to pass on recognized and directly usable knowledge but rather to demonstrate how this knowledge is discovered' (Rüegg, 2004, p 5; Clark, 2006) became institutionalized during the 19th century and at its end dominated the European, US–American and Japanese university landscape. This idea of a university that is largely free from state intervention and whose teaching tasks are linked to scientific research could be implemented only because nation states turned into the major authority responsible for universities and became secularized as well as bureaucratized in the same period.

The authority of the nation state over universities grew because during the 19th century, most European universities faced increasing financial difficulties (Oxford and Cambridge excepted), and nation states began to heavily contribute to the funding of universities (Gerbod, 2004a, pp 85–7; Gerbod, 2004b). In the same period, states created ministries of public education to fund and govern universities (Gerbod, 2004a, p 88–90). The governance of universities included the establishment of national standards for professional training, primarily in medicine, law and teacher education (Gerbod, 2004a, p 90). In the second half of the 19th century, industry began to provide funding for universities as a basis for the emerging university–industry collaboration, especially in Germany and France (Gerbod, 2004b, p 113).

As demonstrated for chemistry, physics and geology, universities created chairs for the newly institutionalized disciplines from the 18th century onwards, which complemented existing chairs in traditional fields like medicine. The systematic teaching of the developing disciplines began in the second half of the 18th century. It was linked to research, and university laboratories became the primary locus of practical education in the sciences and of research in the 19th century. With the growth and specialization of knowledge production, it became increasingly difficult for self-taught part-time scholars to make relevant contributions. At the same time, the growth of knowledge extended the demand for university teachers and university laboratories for empirical research, which created employment opportunities. These processes led to the institutionalization of academic careers and crowded out amateurs, albeit at different rates and with significant variation between regions (Morrell, 1990).

In contrast to the arts, the institutionalization of novelty-driven production in the sciences was based on two transformations, which occurred in two overlapping but distinguishable phases. The first transformation concerned the purposes and practices of producers of scientific knowledge. In the initial scientific revolution, the domains about which knowledge was produced and the understanding of what constitutes an explanation of phenomena became transformed. New practices of producing and communicating contributions became established and existing practices were repurposed. Practitioners began to perceive themselves as building on each other's contributions in jointly stepwise advancing a shared body of knowledge.

The second transformation led to sustained specialized activities of scientific practitioners who formed communities in which they addressed shared epistemic problems with shared approaches. This second scientific revolution led to the emergence of the modern system of scientific disciplines with shared standards for the relevance and quality of contributions. Some of these disciplines developed as 'rapid discovery sciences' in which discoveries are generated by continuously improved sets of embodied practices or research technologies that produce reliable and repeatable results for colleagues/competitors to use as they investigate new phenomena (Collins, 1998, pp 532–8).

These transformations occurred in changing national social contexts of scientific production in which the major organizational features of modern science – the modern research university, state funding and governance of universities and research institutes, industrial research and development – became institutionalized. The social context continued to develop and included the institutionalization of competitive project funding by the state as a major governance change in the second half of the 20th century.

Modern art worlds and science worlds as socio-economic contexts of novelty-driven cultural production

As summarized in the preceding sections, the institutionalization of the arts and sciences as novelty-driven fields of cultural production entailed major changes in: (a) the purposes and practices of novelty production, (b) the relationships between practitioners, (c) the socio-economic contexts that affected novelty production and (d) the relationships between practitioners and these contexts. The initial situations and the trajectories of these institutionalization processes varied considerably between the arts and the sciences but led to comparable structural arrangements.

In the arts they involved the transformation of these production relationships and practices from a service performed for clients into a self-directed consideration of aesthetic problems and questions by members of largely autonomous artistic communities. This transformation was encouraged by the significant socio-economic changes of the period, which included the replacement of patrons and clients by markets as the dominant sources of funding for the arts and a significant expansion and qualitative enhancement of audiences who were interested in, and capable of, appreciating art works.

In contrast, the institutionalization of the sciences as novelty-driven fields involved the transformation of existing practices and conceptions of knowledge production into the production of contributions to a specialized body of knowledge about phenomena that were understood to be the same for all practitioners. This process included the differentiation of knowledge production in separate domains with distinct bodies of knowledge advanced by practitioner communities, a process that has continued ever since and constantly lets new scientific specialties emerge. Patronage by royal courts, the church and wealthy individuals was replaced by patronage from the state and the institutionalization of an associated set of organizations such as modern research universities and funding organizations.

Practitioner communities and their socio-economic contexts have continued to develop but their interrelationships have remained substantially stable for the period of modern novelty-driven artistic and scientific production in much of the 19th and 20th centuries. In this section we outline an analysis of the modern socio-economic contexts of novelty-driven cultural production as the basis for constructing a framework comparing innovation development in different arts and sciences. Our analysis develops ideas from Becker's (1982) conceptualizations of art worlds and Bourdieu's (1975, 1993, 2004) analyses of artistic and scientific fields. The combination of the two approaches is necessary because the comparative analysis of the impact of authority relations on the development of different kinds of innovations requires attention to both changing creative practices (a strength of Becker's perspective) and the different types of actors involved in their development and to the authority of these actors (a strength of Bourdieu's approach).

An analysis of the socio-economic contexts of modern novelty-driven production can be developed from Becker's concept of an art world and its focus on the practices involved in creating art works. For comparative purposes, this approach needs to be further elaborated by (a) extending it to the sciences, for which similarly structured science worlds can be identified, and (b) identifying the main social positions and associated roles of members of art worlds and science worlds. The latter specification of the structure of art worlds and science worlds is informed here by Bourdieu's analyses of artistic and scientific fields, particularly the importance of the distribution of authority between actors involved in the production and distribution of art works and scientific knowledge.

While creative practitioners are central to the production of novelties in the modern arts and sciences, this production would not be possible without support from the specific societal contexts in which it occurs. Actors in these contexts provide resources for the production of novelties and participate in the construction of novelties as relevant contributions. As the reconstruction of institutionalization processes in the preceding sections demonstrated, these socio-economic contexts consist of heterogeneous types of actors including individual as well as various types of collective actors and a variety of procedures and conventions governing the coordination of interactions between them, which occurs through markets, networks, organizations and communities.

Over recent decades, the social sciences have recognized the importance of these heterogenous socio-economic contexts for social change and developed several concepts to capture their impact on processes of change, such as those considered by Zietsma et al (2016). Prominent examples of field concepts, for instance, include the organizational field, which consist of 'those organizations that, in the aggregate, constitute a recognized area of institutional life: key suppliers, resource and product consumers, regulatory agencies, and other organizations that produce similar services or products' (DiMaggio and Powell, 1983, p 148) and the strategic action field, which has been defined as:

> a constructed meso-level social order in which actors (who can be individual or collective) are attuned to and interact with one another on the basis of shared (which is not to say consensual) understandings about the purposes of the field, relationships to others in the field (including who has power and why), and the rules governing legitimate action in the field. (Fligstein and McAdam, 2012, p 9)

These approaches' primary concern is the development and organization of fields rather than the role of fields as socio-economic contexts of production, valuation and change. As such, they are of limited value to the study of changing practices of novelty production in the arts and sciences. The same applies to Caves' (2000) analysis of the 'creative industries', which he applies to the arts and which could be extended to the sciences. His major interest focuses on the organization of collaborations between creative artists and partners in more 'humdrum' industries in producing and distributing new works (Caves, 2000, pp 1–2), particularly contractual relationships, which does not provide the basis for an analysis of changes in artistic and scientific practices through innovations.

Bourdieu's field theory appears to be an ideal source for a comparative framework examining change in artistic and scientific fields because it has already been used in analyses of the arts and the science. He considers fields to be relatively autonomous spheres that are characterized by a distinct logic, and whose members compete for specific kinds of capital according to that logic. A field is a 'space of objective relations between positions defined by their rank in the distribution of competing powers or species of capital' (Bourdieu and Wacquant, 1992, p 113). Fields feature specific logics and sorts of capital but have in common that the purpose of actions is the struggle for dominance through the accumulation of different kinds of capital.

However, this focus on objective relations between actors and their reproduction through the struggle for positions of dominance implies a relative neglect of how practices of novelty production change in different kinds of artistic and scientific fields and makes Bourdieu's somewhat reductive approach less suitable for the comparative analysis of innovation processes in novelty-driven fields of cultural production. In Becker's words, this kind of field theory is concerned with the question of '[w]ho dominates whom, using what strategies and resources, with what results' (Becker and Pessin, 2006, p 285). In contrast, a comparative framework for the analysis of innovations must foreground changing practices of novelty production and treat the distribution of authority in fields as an important condition for the development of innovations rather than the phenomenon whose reproduction is to be explained.

Becker's concept 'art world' provides a more promising starting point for the development of such a comparative framework because its focus is on practices involved in the creation of art works (Becker, 1982). Becker defined an art world as consisting 'of all the people whose activities are necessary to the production of the characteristic works which that world, and perhaps others as well, define as art' (Becker, 1982, p 34). He observed that this definition makes it impossible to draw sharp boundaries of an art world and suggested instead to:

> look for groups of people who cooperate to produce things that they, at least, call art; having found them, we look for other people who are also necessary to that production, gradually building up as complete a picture as we can of the entire cooperating network that radiates out from the work in question. (Becker, 1982, p 35)

Becker's concept and his concern with identifying relevant members of art worlds emphasize the collective construction of art works and the importance of non-practitioner roles in the production and recognition of artworks. It includes a range of actors and their roles according to their impact on the production and appreciation of novelties. The minimal description of its social structure, which is essentially reduced to the analysis of conventions, enables us to suggest a further conceptual development towards a comparative framework by incorporating Bourdieu's ideas of dominance structures into the analysis of the production of art works. The structure of an art world can be differentiated in terms of social positions, types of actors taking up these positions, roles associated with these positions and relationships between actors. This perspective is central to Bourdieu's approach, which focuses on the positions taken by actors and the relationships between actors.

While art worlds as such have not enjoyed much systematic attention in subsequent research on artistic change following Becker's proposal, specific actors in art worlds and their influence on artistic change have been extensively explored. This literature (for example, Wolff, 1993; Jensen, 1994; Woodmansee, 1994; Fitzgerald, 1995; Lopes, 2002; Baumann, 2007; Cottington, 2022) suggests that three general types of roles can be distinguished according to their functions in the construction, recognition and experience of art works: creative producers, audiences and intermediaries.

The role of a creative practitioner in the modern arts is to produce novel contributions that address collective aesthetic problems and questions. It is thus linked to the membership role in a community whose collective identity is based on members' common understanding that they are addressing shared aesthetic problems. Artists can be members of more than one community simultaneously, which in turn means that practitioner communities overlap. Artists can also take roles other than that of a creative practitioner, as did writers in France commenting on Impressionist paintings in the 1870s and 1880s, acting as interpretative intermediaries in this art world (Cottington, 2022, pp 110–8).

The second main role is that of an audience member experiencing art works. Audiences play a major role in the production of novel art works because they not only contribute to the recognition and valuation of art works but also pay for opportunities to experience art works and thus significantly contribute to the funding of most art worlds in general and of artists in particular. Again, membership of an audience is a particular role, which may be taken in addition to other roles. For example, it is not uncommon for artists to take audience roles for each other.

The third major role in art worlds is that of an intermediary who mediates between practitioner communities and their audiences. This mediation is necessary for several reasons. In several of the arts, such as the performing and cinematic arts, the production of art works itself requires significant resources and must be organized as a collective process. This means that funds must been raised (usually at least partly from audiences) and must be invested in collective work processes. These roles are usually taken by organizations such as theatres, film companies and publishers.

Additionally, opportunities for audiences to experience art works must be organized in all art worlds. This includes informing audiences about such opportunities through a variety of media and the organization of exhibitions of art works, of performances or of the distribution of art works to different kinds of audiences. Finally, some audiences must be supported in experiencing art works through commentary, framing and explanation. Some of these intermediary roles are taken by individuals, such as the roles of critics, while others are taken by organizations such as publishers and film companies.

These three sets of roles of a creative practitioner, audience member and intermediary form the basic structure of art worlds. Although executing these roles requires different competences, members of art worlds can, and often do, take more than one role in the production and appreciation of art works. In executing these roles, members of art worlds contribute to the funding, production, distribution and appreciation of art works and exercise authority over these processes to varying degrees.

The idea of art worlds in which actors in a variety of different roles contribute to the construction, recognition and valuation of art works can be extended to the sciences. There is little support for this extension from science studies because previous applications of the social-world approach to the sciences have equated science worlds with practitioner communities (Gerson, 1983; Star and Gerson, 1987; Fujimura, 1988; Clarke, 1990), thereby depriving themselves of the opportunity to study the participation of heterogeneous actors in the production and appreciation of scientific novelties. The only exception to this narrow conceptualization of science worlds was Simon who, similarly to Becker, 'find[s] the concept of science worlds useful for talking about the myriad of actors other than professional scientists whose participation is intrinsic to the production of natural-scientific knowledge' (Simon, 2002, p 225) but does not develop the concept.

Similar to art worlds, the relevant socio-economic contexts in which scientific contributions are produced and appreciated can be understood as science worlds consisting of practitioner communities, audiences and intermediaries mediating between the two. The major difference between modern art worlds and science worlds is that scientific practitioner communities constitute their own main audience. Scientific practitioners address their contributions to each other because they collectively advance a shared body of knowledge and use others' contributions in the production of their own. Therefore, the appreciation of scientific contributions focuses mostly on their utility for further work, and only fellow practitioners are able to make that assessment.

We now turn to a brief discussion of the different kinds of these creative practitioner, audience and intermediary roles that have developed and changed in different modern art worlds and science worlds. Considering first the role of creative practitioner, this role in the modern arts and sciences is also a membership role in a community of creative practitioners. Communities of creative practitioners in these art and science worlds have in common a collective identity based on their members' perception of being engaged with shared aesthetic/epistemic problems. Members are also aware that other members' creative contributions addressing these problems are relevant to their own production. In their work, members of practitioner communities share resources and adhere to collective standards of relevance and quality of contributions. Therefore, practitioner communities in the arts and sciences are not only communities of practice (Lave and Wenger, 1991) whose members are engaged in the same practices but also maintain various kinds of productive relationships with each other when conducting their work.

The formation of practitioner communities addressing shared aesthetic/epistemic problems was a constitutive process in the institutionalization of the arts and sciences as novelty-driven fields of cultural production. However, they have had unequal attention in analyses of the modern arts and sciences. While the study of scientific communities can be traced back to Fleck (1935), their investigation became a focus of research in the sociology of science after Kuhn (1970 [1962]) emphasized their role as units of collective knowledge production. Although such interest waned with the microsociological turn in the sociology of science at the beginning of the 1980s, they have continued to be implicitly considered as the primary social context of the production of scientific knowledge and to be defined as the core unit of analysis of field-comparative science studies (Whitley, 2000; Gläser, 2006).

In the history and sociology of art, the investigation of practitioner communities and their varied kinds of productive relationships has had a less prominent position. While many studies have considered the role of collective actors in the arts, they mainly focused on temporary collective associations that brought about particular aesthetic changes – artistic movements – and disregarded the more stable communities of practitioners in which these changes occurred. More recently, though, Cottington has emphasized the importance of a shared identity for the Parisian avant-garde as an artistic community:

> [while] the two factors of aesthetic innovation and independent lineage could be described as necessary but not sufficient for the consolidation of the Parisian avant-garde … what was crucial in securing that consolidation, and enabling their dissemination across the formation, was a third, namely the collective identity – the mutual recognition among members of this community of their common and distinctive

> artistic aspirations – that they helped to establish in the two decades around 1900. (Cottington, 2022, p 142)

Beyond this recognition of an identity-based community whose members are aware that they address the same aesthetic problems, there has been little systematic treatment of how communities of artists jointly produce novelties and change.

Second, membership of an audience contributes to the recognition of art works and scientific knowledge and in many cases to the funding of artistic and scientific production. Audiences are thus members of art worlds according to Becker's definition. Becker confirms this when he discusses the role of conventions for establishing audiences and delineating art worlds:

> Conventions known to all well-socialized members of a society make possible some of the most basic and important forms of cooperation characteristic of an art world. Most important, they allow people who have little or no formal acquaintance with or training in the art to participate as audience members – to listen to music, read books, attend films or plays, and get something from them. Knowledge of these conventions defines the outer perimeter of an art world, indicating potential audience members, of whom no special knowledge can be expected. Art forms designed to reach the maximum number of people in a society take most advantage of these resources. Some – popular television shows, for instance – do that in an established way. (Becker, 1982, p 46)

Audiences vary in their size, knowledge of conventions and investment in the skills that are necessary to appreciate the works of art that are characteristic for an art world. In the previous quote, Becker describes the largest possible audiences, namely mass heterocultural audiences who only need to be familiar with general societal conventions to appreciate some art works. In other art worlds, audiences can be small, and the audience role can be (and often is) executed primarily by other artists from the same or another community. As Charles Rosen emphasized in his discussion of *The Irrelevance of Serious Music*:

> The life of music is based not so much on those who want to listen, but on those who want to play and sing … Musicians need and want an audience from time to time, but the public concert is only a small part of the making of music, not the whole of it: playing for a few friends, playing with other musicians, and even playing for oneself still provide the foundation of musical life. (Rosen, 2000, p 296)

Other works produced for much larger audiences depend on smaller groups of aesthetically informed audiences for recognition as works of art. Art works whose appreciation requires an investment in developing such skills are usually addressed to knowledgeable audiences or what Bourdieu (1993, pp 115–25) termed 'restricted audiences'. The size of an audience might be quite small if the audience role includes the acquisition of property rights of an art work, which occurs when art works are privately commissioned or bought by collectors.

In the sciences, creative practitioners constitute the most important audience because they are often the only ones who are competent to recognize a novelty as a relevant contribution to the community's production of knowledge. However, practitioner communities in the sciences, social sciences and humanities are not without non-practitioner audiences, which vary in their ability to assess creative contributions. These include other practitioner communities whose epistemic concerns may be more or less distant and professions which receive the same initial academic education as those who later become members of scientific communities (Whitley, 1985). The widest audience consists of the societies in which they are embedded through practitioners' employment in organizations and the funding of their research. While this audience is largely unable to appreciate particular novelties produced in scientific communities due to its limited expertise, the acknowledgment that they are produced, and that knowledge about the world is extended and informs teaching, supports the public funding of research.

The focus of the social sciences and humanities on societal processes additionally provides their producers with more specialist audiences of those involved in these processes. Finally, scientific communities dealing with particular societal problems (the technology and policy relevant natural sciences and social sciences) have specialist audiences. As in the arts, these audiences can be very small, for example in the case of commissioned research (which is common in academy–industry collaborations). An interesting split occurs in the case of patenting, where the publication of findings in patents makes them accessible for observation and appreciation to large audiences but limits their use to patent holders and their licensees.

Third, there are at least three kinds of intermediary roles that can exercise some authority over aesthetic/epistemic practices, standards and purposes: resource providers supporting the production of contributions, distributive roles and interpretative roles. Resource provision roles organize and allocate resources for their production and diffusion, distributive roles display and diffuse novelties to audiences in different media and channels, while interpretative roles analyse and evaluate novelties for different kinds of audiences in a variety of ways. While these roles may be performed by different actors in different fields and socio-economic contexts, quite often groups and organizations combine them, as when museum curators organize exhibitions of paintings to present a particular interpretation of aesthetic change or when film production companies fund and distribute new films for particular markets.

Considering first contributors supporting the production of art works, this may depend on significant and sustained material investments or on contributions by other practitioners. Examples of intermediaries supporting the production of art works include film companies or publishers of literature and music as well as impresarios who perform a similar role in the stage arts. These intermediaries receive returns on their investment from the sale of art works or access to them to the audiences. Most but not all of them are commercially oriented and need to make a profit. Their authority in art worlds is usually high because without them the art works would not exist or would not reach their audiences.

In addition to these non-creative actions and material investments in the production of art works, co-producing intermediaries may also make creative contributions. This occurs when practitioners from other artistic communities contribute to producing an art work. For example, an intermediary role as supporting the production of an art work can be ascribed to practitioners of the performing arts who simultaneously execute roles of creative practitioners and intermediary roles of an art work.

A second important intermediary role is that of making art works accessible to audiences by distributing or displaying them. Intermediaries supporting the production of art works are often also involved in their distribution, as in multiply reproduced literary and musical works. Distributing art works to audiences can also be a separate role, as is the case for art exhibitions, museums and galleries as well as the curators they employ.

Interpretative intermediaries differ between knowledgeable critics addressing educated and interested audiences of non-practitioners in specialist reviews, on the one hand, and less well-informed commentators principally concerned to attract the attention and interest of larger and more heterogenous audiences through a variety of channels, on the other hand. The former are more likely to invest time in learning about the aesthetic/epistemic merits of particular discontinuous novelties for art worlds' and science worlds' purposes and support practitioners developing them into innovations than are more commercially focused intermediaries. These are usually more reluctant to recognize and support novelties that challenge established conventions as serious works of art.

In her analysis of the professionalization of art criticism in Victorian Britain, Elizabeth Prettejohn (1997) has highlighted the growing divergence of periodical and newspaper art criticism between generalist and professional critics and audiences:

> By the 1870s, there was a sharp distinction between professional art critics, whose names and credentials were known in public, and

> generalists who remained anonymous … Dramatically different kinds of critical writing were offered to different readerships in the later Victorian period. Not only specialist art periodicals, but general periodicals aimed at an upper middle-class readership tended to employ professionals. However, mass-circulation newspapers and weeklies continued to employ generalists. (Prettejohn, 1997, p 79)

While the latter consistently privileged the picture's emotional impact over its formal or visual qualities, the professional critics preferred to concentrate on art-historical stylistic comparison:

> In the literature of art history, stylistic comparison was used to solve problems of attribution … The professional critics altered the function of comparison, using it to locate contemporary British art within their own sphere of professional expertise, the transhistorical and cross-cultural sphere of 'art great in the abstract'. Technical vocabulary and stylistic comparison established an esoteric descriptive language, presented as applicable to all kinds of visual art, but inaccessible to the generalist critic. (Prettejohn, 1997, p 85)

Similarly to the arts, interactions between scientific communities and their varied audiences and clients are facilitated by intermediaries executing specialized roles. The production of novelties is supported by employment organizations such as universities and research institutes, which provide researchers with salaries, material resources for research and access to the knowledge of their scientific communities. Employment organizations also provide the channels through which resources from external funding organizations are accessed and utilized for the production of scientific novelties.

The role of distributing novelties to creative practitioners' main audience – their own scientific community – is executed by publishers of scientific journals and books. More recently, platforms for electronic publications, particularly preprint servers, have begun to take this role, too. These intermediary roles are often shared or combined with roles taken by practitioners themselves, who manage scientific journals and book series and organize conferences or workshops. Intermediary roles for the distribution of scientific knowledge to non-practitioner audiences like commercial and non-commercial users, science policy, or the wider public include publishers of professional journals, professionals interacting with mass media, teachers and researchers themselves.

By contributing to the production of art works and scientific findings and by mediating resource flows, access to products and their appreciation, incumbents of intermediary roles perform necessary functions for the novelty-driven production in modern art worlds and science worlds. Creative practitioner roles, audience roles and intermediary roles are productive roles in art worlds and science worlds in that they are necessary for the production and appreciation of contributions addressing the shared aesthetic/epistemic problems of an art or science world. Understanding novelty-driven production in the arts and sciences and its change requires the analysis of the structure of modern art worlds and science worlds, whose emergence and current main components were the subject of this chapter.

In the following chapters, we analyse how the social structures of art worlds and science worlds affect the development of different kinds of innovations in the modern arts and sciences. These kinds of innovations and the specific aesthetic/epistemic change they produce will be identified in the next chapter. Their development is shaped by the productive relationships creative practitioners maintain in their novelty-driven production and by the distribution of authority between the productive roles in art worlds and science worlds, which we will discuss in Chapter 4. In Chapter 5 we present nine case studies of the successful development of different kinds of innovations in differently structured art and science worlds. How their development and acceptance depended on a variety of combinations of practitioner actions, shifts in intermediary roles and priorities, and changes in the nature and priorities of audiences is discussed in Chapter 6, together with a brief consideration of two recent changes in the socio-economic conditions governing authority relationships in art and science worlds.

3

Varieties of Innovation and Change in Novelty-Driven Fields

In his 1975 article criticizing economists' approach to analysing technological change Nathan Rosenberg suggested that many of the problems he identified stemmed from their acceptance of Schumpeter's sharp distinction between invention, innovation and imitation. In particular, they have continued to assume that:

> Inventions come onto the Schumpeterian stage already full-grown, and not as objects or processes the development of which is a matter of explicit interest; nor are subsequent improvements or modifications of the invention typically treated as significant. As a result, invention appears as an isolated phenomenon … and the stages through which inventions proceed on the way to full commercial application and exploitation, never emerge [as a subject of analysis]. (Rosenberg, 1975, pp 463–4)

In the five decades since this paper was published, the field of innovation studies has expanded considerably to include analyses of innovations in the increasingly important services industries (Miozzo and Grimshaw, 2006; Miles, 2016) as well as numerous comparisons of regional and national innovation systems (Lundvall, 1992; Nelson, 1993; Braczyk et al, 2003), but the lacunae identified by Rosenberg remain under-researched (Godin, 2012, 2016; Martin, 2016).

These include the processes by which inventions become transformed into technically feasible products or processes that can be implemented on a large scale and subsequently introduced successfully to the market. While much attention has been paid to how firms commercialize innovations more or less effectively in what Godin (2012) terms the second tradition of technological change research, rather less effort has been expended on understanding the processes by which new inventions are modified and elaborated to become effective innovations. Additionally, the role of complementary inputs and other resources in successful commercialization of new products and processes has tended to be underexamined, as has the role of users in designing and improving new technologies despite von Hippel's (1976) emphasis on the importance of researchers actively influencing the development of new scientific instruments and Rogers (2003) highlighting the importance of 'learning by using' during the diffusion of innovations for further improving them. Overall, since the field of innovation studies emerged several decades ago, most empirical studies have tended to focus on a relatively narrow range of innovation types, as Martin (2016, p 434) has pointed out.

In an extension of his critical analysis of economists' conceptualization of technological change, Rosenberg (1975, pp 474–9) suggests the same points could be applied to the development of intellectual innovations in the social sciences. In particular, he highlighted the need for new ideas and techniques in economics to be subjected to sustained critical analysis and conceptual development if they were to be operationalized in ways that enabled them to be used by other practitioners to make significant contributions. Such operationalization is crucial if novelties are to become established as significant contributions that change colleagues' practices and purposes. Additionally, in many cases of radically unconventional approaches, a number of complementary changes in research procedures and intellectual resources have been critical in ensuring their widespread adoption and development, such as the construction and diffusion of reliable data sets and the availability of computer technologies for dealing with large-scale analyses.

A third factor affecting the development of novel approaches and practices into innovations that change the direction of a social-scientific field concerns shifts in dominant socio-economic priorities and support, such as the increasing concern with unemployment in the 1930s affecting the level of economists' interest in Keynes' *General Theory*. Finally, as in most technological innovations, the processes of diffusion and adoption of conceptual and practical innovations in the social sciences usually modify and further develop the nature and usefulness of intellectual changes, in both the field of initial articulation and use and other areas of research that often deal with different problems.

Similar points can be made about the development, modification, diffusion and use of innovations in other sciences and in the modern arts. In these novelty-driven fields of cultural production, the creation of new ideas, practices and problems is an everyday activity that produces an enormous variety of novelties (Caves, 2000). For these to become innovations that change established purposes and practices, they have to be considered useful in dealing with current aesthetic and epistemic priorities and problems and further developed by competent practitioners in the production of their own contributions to collective purposes. Innovations in the modern arts and sciences are, then, processes that alter the nature of fellow practitioners' production practices and purposes on a continuing basis.

Types of innovation and change in novelty-driven fields

These processes of exploring and modifying novelties in the development of productive innovations in novelty-driven fields of cultural production vary in the kinds of practices, conventions and priorities that change and in the degree to which they establish new approaches and goals that deviate significantly from current purposes and standards. While all novelties in these fields are, by definition, different in some way from previous outputs, both the extent and kind of innovation processes they trigger and the aesthetic/epistemic change they generate vary considerably between fields and circumstances.

A common contrast between types of change in the modern arts and sciences, for instance, sharply distinguishes between the minor variations in artistic and scientific practices dealing with everyday puzzles characteristic of what Kuhn termed 'normal' science from the radically disruptive nature of a 'revolutionary' innovation with its transformation of dominant understandings of a field's problems, concepts and appropriate approaches for dealing with them (Kuhn, 1970 [1962], pp 36–70, 92–110; Clignet, 1985, pp 2–6). However, many of the attempts to apply Kuhn's summary of the differences between the incremental shifts in puzzle-solving activities characteristic of normal science and the transformative nature of paradigm change in scientific revolutions to changes in the social sciences, humanities and the arts overemphasize the contrast between minor changes in the former and the wholesale transformation of fundamental concepts and standards in the latter.

As Clignet (1985, pp 72–4) concluded in his discussion of revolutionary changes in the arts, most, if not all, discontinuous changes in artistic practices and problems are partial rather than transformative of entire artistic communities' understandings of fundamental concepts, questions and techniques. Just as some changes in the nature of phenomena studied and problems addressed in the sciences are more akin to Mulkay's (1975) account of significant innovations as the opening up of new areas of ignorance rather than the total reformulation of dominant goals and standards, so too some discontinuous innovations in the arts and humanities can become legitimized as additions to established practices and approaches that do not completely transform them.

Instead of reducing the variety of innovations and change in the modern arts and sciences to the simple dichotomy of continuous incremental shifts in problems and approaches, on the one hand, and discontinuous transformations of dominant questions and epistemic/aesthetic standards, on the other hand, then, we need to consider how different innovations generate more varied kinds of change. In particular, the nature and scope of what has changed, such as the major questions and concepts of the field, dominant aesthetic/epistemic standards, recognized ways of dealing with problems and evaluating contributions and novelty production practices need to be specified. Additionally, the extent of collective commitment to the change throughout practitioner communities and their associated intermediaries has to be considered.

These characteristics of different kinds of artistic and scientific innovations and change can be summarized in terms of two general differentiating dimensions. *Variations in the nature of discontinuous changes* in practices, methods, questions and constitutive concerns and approaches can be distinguished. These can be summarized as three sets of changes. First, changes in the accepted ways of dealing with the field's problems and questions that can encompass changes in the standards governing the production of contributions to solving common problems as well as incremental shifts in everyday practices. They depart from established ways of producing novelties but do not usually modify the nature of problems and questions of the field. Second, changes in the nature of problems and questions considered legitimate and significant concerns of the field, which can involve changes in acceptable ways of dealing with these. Third, changes in the field's constitutive problems and questions that imply substantial changes in competent practitioners' understandings of the field's dominant concerns. These often involve changes in novelty-production practices.

The other dimension concerns *the extent to which members of the field implement these changes* in their production and assessment of contributions to collective purposes. This refers to the level of collective commitments to these new approaches and practices, which can vary between limited groups of practitioners developing incremental shifts in everyday practices of novelty production to the rapid and widespread commitment of a producer community to new conceptualizations of a field's constitutive concerns and ways of dealing with them.

Combining these characteristics of innovations and changes in novelty production, distribution and evaluation in the modern arts and sciences enables us to distinguish four major types: Drift, Process, Complementary and Displacement. In its focus on the processes triggered by particular novelties rather than on characteristics of these novelties, this typology differs from distinctions between radical and incremental innovations in the literature on technological innovations (Tushman and Anderson, 1986; Teece, 2010), many case studies in the history and sociology of the arts and sciences (Pinch, 1986; Cozzens, 1989; Bourdieu, 2017), and attempts to measure the novelty or originality of scientific contributions focusing on how innovation-triggering novelties differ from the current state of the art

(Shibayama, 2020; Matsumoto, 2021). These alternative approaches are often focused on how (or in the case of indicators, how much) a particular novelty deviates from current practices, which implicitly casts an event – the communication of a particular novelty – as the innovation. In contrast, the typology presented here is based on understanding innovations as processes that are triggered by such novelties, and distinguishes innovations according to the content and extent of lasting change in practitioners' purposes and practices that are brought about by innovation processes.

The major features of these kinds of innovations are summarized in Table 3.1 with some examples in the arts and sciences. Those underlined will be discussed in more detail in this and later chapters. First, **Drift innovations** modify dominant practices in a field in a largely incremental manner that enables competent practitioners to continue producing contributions to collective goals in a substantially additive process that does not require the rejection of current approaches and standards or major investments in learning new practices to deal with new kinds of questions and reorganizing patterns of collaboration. These can be distinguished from changes occurring as a

Table 3.1: Types of innovation and change in novelty-driven fields of cultural production

Characteristics and examples	Drift	Process	Complementary	Displacement
Nature of discontinuous change	None	Establishing new methods and approaches for dealing with common problems	Adding new questions and approaches to the canon, often accompanied by new methods for dealing with them	Reconceptualization of a field's constitutive concerns and approaches
Level of producer collective commitment to changes	Varies	Varies	Limited	High
Examples in the arts	Performance practices	*Plein air* painting, multiple divergent narratives in fiction	Metafiction in Anglophone fiction, Impressionism in painting, *Nouvelle Vague* in French cinema	Cubism in painting and sculpture, Bebop and modern jazz
Examples in the sciences	Minor technical improvements	Realization of Bose-Einstein Condensation (BEC) up to 2010s, Radio Astronomy	BEC from 2010s, Ethnomethodology, Evolutionary-Developmental Biology (Evo-devo)	Oncogene Approach, Financial Economics, Plate Tectonics

Note: Underlining indicates that the innovation is the topic of a case study in Chapter 5.

result of the inevitable variations in everyday replications of standard practices and models. As Kubler (1962, pp 54, 62–70) has stressed in his discussion of linguistic change, such small variations are not only inevitable but may also accumulate to important long-term changes in everyday production practices. In contrast, drift innovations here result from more selected shifts in everyday practices of novelty production that promise to create more interest and/or improve the value of contributions to collective purposes, such as performance changes in the arts or minor modifications to technical procedures in the sciences.

These innovations combine relatively low investment costs in learning new practices and accepting minor modifications of the ways that aesthetic/epistemic problems are understood and assessed with limited changes to previous patterns of work organization and control. As Becker puts it in his account of these kinds of change in art worlds: 'No one's livelihood is threatened. The people who cooperate to produce works will continue to do so, even though the work they produce is different' (Becker, 1982, p 304). They do not disrupt the standards governing practices, approaches and dominant collective understandings of the field's major problems and questions.

The development of the novelties generated by participants in these novelty-driven fields into drift innovations that are adopted by producers is often a matter of choice for the members of the field as a whole, and hence not obligatory for all to incorporate into their everyday practices. A limited scope for developing drift innovations is quite common in the performing arts, as in the case of string players using vibrato to heighten tension and generate an emotional response in a period when it was unconventional (Meyer, 1956, p 66). However, when such shifts seem likely to improve the chances of producing significant contributions to important collective purposes, initiating a more widespread snowballing pattern of adoption, they can rapidly become more obligatory for practitioners to use as necessary components of their competences. In some cases, such extensions of producers' capabilities are reinforced by audiences' demand for new experiences, as in the development and popularity of aesthetically novel jazz songs and those derived from musical comedies that become established as core parts of the repertoire (Becker, 1982, p 60).

Second, **Process innovations** establish discontinuous ways of creating novelties contributing to the field's problems. They involve developing and applying new production procedures and approaches to problem solving for the creation of collectively valued outputs that transform or significantly enhance rather than incrementally improving current skills. In promoting *plein air* modes of painting landscapes in France that involved substantial periods of sketching and painting in the open air, for example, the Barbizon group sought to establish a more 'natural' and accurate representation of the French countryside than the prevalent neoclassical styles favoured by the Academy. They produced less overtly polished works with more visible brushstrokes, varied effects of light and paid greater attention to the sky than was usually expected (Boime, 1971, pp 147–8; Malafronte, 2009). As Fraser (2017, p 7) observes: 'Barbizon works were quickly executed using broad brushstrokes. Their colours were taken directly from the landscape itself, rather than a particular palette, so as to capture a specific moment or play of light'.

While such changes could sometimes be achieved with relatively low investments of time and other resources, in other cases the development of new technical practices and approaches took several years and required access to major instrumentation and laboratory resources, such as those involved in the creation of Radio Astronomy and the experimental realization of Bose-Einstein Condensation (BEC) in low temperature physics. In the case of Radio Astronomy, its initial development followed the discovery of cosmic radiation in radio telephony in 1937 and the wartime improvement of radar physics instruments in Australia and the UK. While early postwar results were acknowledged by some optical astronomers as potentially solving major difficulties in astronomical research, their lack of precision in the positioning of radio sources limited the epistemic usefulness of these techniques. In the late 1940s and 1950s, though, they were greatly improved by the construction of new 'radio telescopes' and allied instruments that were developed with the support of considerable government grants. By the mid-1950s the observations generated by these methods had become widely accepted by the international astronomical community to the extent that in 1958 the International Astronomical Union changed its approach to galactic coordinates and the definition of galactic pole in response to radioastronomical findings (Edge and Mulkay, 1976; Munns, 2013; Chapter 5).

BEC occur when gases of atoms are cooled to temperatures very close to absolute zero and particles lose their individual identities, coalescing into a single blob. Although widely accepted as a theoretical phenomenon, the experimental realization of BEC was regarded by many physicists as being very difficult, if not impossible, to achieve for both theoretical and technological reasons. It was finally accomplished by two US groups in 1995 after many years spent constructing the necessary experimental system and understanding how it worked (Laudel et al, 2014b; Whitley et al, 2018; Chapter 5).

Gaining adoption and further development by other producers may also take some time, even in fields with high levels of consensus on technical practices. Despite the BEC achievement being widely recognized as a major achievement, the scientific community was initially undecided whether this would be the end of a long quest or whether it would open up new research opportunities. Although the initial success of creating BECs in cold atom gases has been followed by a number of attempts to create BECs of different atoms, as well as photons, solitons and other physical entities, researchers have continued to experience experimental difficulties, as in the cases of atoms that are not susceptible to laser cooling.

More importantly, however, BECs turned out to be very useful experimental platforms for a wide range of fundamental questions of quantum mechanics such as quantum many-body phenomena. According to Fallani and Kastberg (2015, p 4): 'A completely new field of research has emerged, in which ultracold atoms are used as elementary components in the investigation of fundamental properties of quantum matter'. Georgescu (2020, p 396) reports that: 'BECs have also been used to create atom lasers, atomic clocks and gravitational, rotational or magnetic sensors with excellent sensitivity'. With the exception of the failures mentioned previously, 'making atomic BECs in the lab has become routine. There are commercial tabletop BEC systems available and there are compact systems that can be used for portable sensors or free-fall experiments' (Georgescu, 2020, p 396). In these respects, the initial BEC process innovations in the late 1990s have developed some characteristics of complementary innovations in the early 2010s.

Third, **Complementary innovations** establish new questions and approaches to be pursued in addition to those currently dominating the field and so enlarge the kinds of problems considered to be important as well as novel ways of dealing with them. These changes do not, though, result in the substantial replacement of the field's dominant questions and conventions by quite different ones, nor do they typically require the bulk of competent practitioners to adopt them. They provide alternative trajectories of creative production and career opportunities to those focused on contributing to the standard purposes of the field, even though their proponents often have to struggle to gain legitimacy among established practitioners, especially when they are seen to threaten their authority and control over access to resources.

Complementary innovations can occur when discontinuous ways of generating unconventional outputs become, after some development, accepted as valid approaches to producing contributions that some practitioners can legitimately adopt without most being obliged to do so. The importance of such complementary developments for the field varies. In the case of US Ethnomethodology, for instance, although critical analyses of the prevalent empirical assumptions and methods of American postwar sociology did generate a separate subfield with its own publications and training groups, this subfield remained small and did not gain broader institutional recognition and adoption across the discipline (Mullins, 1973; Hammersley, 2018).

In other cases, producing novelties that deviate from dominant conceptions of a field's purposes and currently acceptable approaches to dealing with them can eventually become acknowledged as useful additions to the repertoire of valid questions and approaches without necessarily transforming authority relationships and the social organization of cultural production. Such acceptance often occurs when they are combined with other innovations and changes in socio-economic contexts, as Kubler (1962) and Rosenberg (1975) have suggested. In the arts, this recognition and incorporation into the repertoire of established practices is often dependent on their ability to attract the support of influential intermediaries and a substantial audience, as in the case of Metafiction in post-Second World War Anglophone literature.

Metafictional novels became a significant literary innovation in the UK and the US in the 1960s and 1970s. Emphasizing the constructed nature of fictional narratives and the self-conscious role of writers in making particular choices in creating them, metafictional novels make the act of literary composition its own subject in opposition to the language of the realist novel. In his short story 'The Magic Poker', for instance, Robert Coover frequently refers to his own authorial role in creating characters and their actions, as in:

> I have brought two sisters to this invented island, and shall, in time, send them home again. I have dressed them and may well chose to undress them. I have given one three marriages, the other none at all, nor is that the end of my beneficence and cruelty. (Coover, 1969, p 11)

Metafictional texts additionally highlight the linguistically constructed nature of fictional narratives and characters. As suggested by Patricia Waugh (1984, p 24) in her account of Metafiction:

> Whereas loss of order for the modernist led to the belief in its recovery at a deeper level of the mind, for metafictional writers the most fundamental assumption is that composing a novel is basically no different from composing or constructing one's 'reality'. Writing itself rather than consciousness becomes the main object of attention. Questioning … the authority of consciousness, of the mind, metafiction establishes the categorization of the world through the arbitrary system of language.

This emphasis on the constitutive role of linguistic conventions constructing not only novels but also everyday activities and reflections on them, such as literary criticism, drew attention to the consequent commonalities between writing novels and constructing critical analyses of them, questioning the ability of critical language to refer objectively and authoritatively to the literary text (Currie, 1995).

Despite the radical claims of many metafictional writers in opposition to established conventions, later in the 20th century many of these techniques became less of a confrontational challenge to tradition than a helpful complement to it (Klinkowitz, 1980, 2017). The development of successful academic careers by creating novel critical positions and techniques in an expanding higher education system, often invoking influential deconstructionist arguments, was combined with the construction of influential writing careers drawing upon these approaches, as in the cases of Ronald Sukenick and Raymond Federman, who completed their PhDs at leading institutions before publishing their first novels (Chapter 5).

Fourth, **Displacement innovations** replace dominant conceptualizations of the central problems of a field and how they should be approached. In many sciences, these become obligatory for practitioners to develop if they are to continue to produce collectively recognized and valued contributions, often to the reconstituted field. They replace dominant ways of working, approaches to conceptualizing problems and the standards governing the evaluation of contributions with substantially new ones.

For example, the molecular biological approach to the study of cancer replaced previous approaches in the late 1970s and 1980s. This approach combined the 'oncogene theory' that 'cancer is caused by normal cellular genes called "proto-oncogenes" that somehow turn into cancer genes, loosely called "oncogenes"' (Fujimura, 1988, p 262) with recently developed molecular biological methods to investigate these genes. Analysts do not fully agree, though, on the nature of the practices and concepts that were replaced by the molecular biology approach.

Historians focusing on the content of theories see an emerging 'oncogene paradigm' either as merely a more precise formulation of the 'regulatory vision' of cancer already in existence (Morange, 1997) or as revolutionary because it transcended a dichotomy between exogenous and endogenous explanations (van Helvoort, 1999). Sociologists emphasize that the 'oncogene theory' – the new model of cancer emergence – was linked to a new set of practices, namely the application of molecular biological methods to the empirical study of cancer (Studer and Chubin, 1980; Fujimura, 1988). All analysts agree, though, that the 'oncogene theory' came to dominate the explanation of cancer and molecular biology concepts and methods dominated empirical research on cancer from the 1980s.

There was little opposition to the new genetic approach to explaining cancer, not least because the search for a cancer virus had not been successful and funding for this search was discontinued in the US in 1978 (Studer and Chubin, 1980, pp 96–9; Fujimura, 1988, pp 271–2; van Helvoort, 1999, pp 319–20). However, turning to the new approach required a significant investment in learning the new molecular biological methods. Originally, the recently developed methods for investigating genetic causes of cancer and mechanisms that turned normal cells into cancer cells were weakly codified and difficult to learn. The quick adoption of the new molecular biological research practices was made possible by their standardization, which took the form of 'pre-fabricated biological materials (reagents, probes), procedural manuals (often called "cookbooks") spelling out "recipes" for action, industrial standards, computing protocols, and instruments that automate many procedures' (Fujimura, 1988, p 266). This standardization drastically reduced the protected space necessary to develop the innovation – the time and resource investments for learning the new methods – and increased the likelihood of success (Chapter 5).

In some cases, displacement innovations not only replace established practices and purposes but also lead to the transformation of fields of cultural production by focusing them on new kinds of problems and practices and establishing novel aesthetic/epistemic conventions and authority relationships. The transformation of business finance research and teaching into Financial Economics, for example, not only changed the nature of the topics considered and ways they were tackled and communicated in the leading American journal, the *Journal of Finance*, between 1960 and 1980, but also altered the backgrounds and competences of their authors, referees and office holders in the American Finance Association (AFA).

From covering a wide range of subjects, including the nature and functioning of institutional arrangements, legal structures and long-term financing of companies and investment projects, practically all of the papers published in the *Journal of Finance* and the other two major journals founded since 1960 – the *Journal of Financial and Quantitative Analysis* and the *Journal of Financial Economics* – relied extensively on mathematical and/or statistical analyses focused on asset valuation models and capital market price movements in highly idealized markets. Whereas practitioners in financial institutions could, and did, contribute to finance journals up to the mid-1960s, by the 1980s they were dominated by academics, as were the officers and directors of the AFA. In effect, research in business finance had been transformed into a branch of neo-classical micro-economics concentrated on highly abstract models of stock price movements (Keenan, 1970; Weston, 1974; Whitley, 1986a; Chapter 5).

In the arts, displacement innovations sometimes become established when radically new approaches that reconceptualized the constitutive problems of a genre are supported by a novel social formation, as suggested by David Cottington (2022) in his account of Cubism in the construction of the Parisian avant-garde in the early 20th century. Cubism was an artistic movement that initiated a pictorial revolution in the ways that images were constructed and presented. Generally understood to have been initiated by Georges Braque and Pablo Picasso between 1907 and 1914, it rejected the traditional techniques of *chiaroscuro*, or using light and shade to create the illusion of volume and depth, single-viewpoint and aerial perspective, foreshortening of people and objects and the common presumption that paintings must present an illusion of reality at a particular time.

Instead, Cubists combined abstraction, multiple viewpoints, distortions of forms, ambiguous spatial relationships and limited use of different colours to emphasize the flat, two-dimensional nature of the canvas and the many and varied ways of seeing and understanding the world (Cox, 2000, pp 11–3; Antliff and Leighten, 2001, pp 7–10). In particular, the early Cubists focused on three distinct tensions or problems in painting: (a) how to represent the three-dimensional nature of objects and people on a two-dimensional canvas without creating a mere illusion of depth; (b) how to resolve the conflicting priorities of form and colour; and (c) how to deal with the tension between representing a moment of experience on the canvas and the extended process of thinking about and constructing images of experience (Baxandall, 1985, pp 44–5).

While such issues had been considered in a variety of ways in European painting over a long period, they were addressed more directly in the work of Courbet, Manet, Van Gogh, Gauguin, Cézanne and others which deviated from the established academic conventions and developed novel approaches in the second half of the 19th century. However, Braque and Picasso challenged the purposes and techniques of artistic realism in a more systematic and extended manner, particularly between 1909 and 1913. Whereas they had previously tended to progress from one substantial project to another, according to the rhythm imposed by annual exhibitions at the public and private Salons:

> After early 1909 [Picasso] made no more big paintings, undertook no more major projects in the salon manner … [and] … adopted a very different way of working, in which each painting was part of an open series, a fresh point of departure for a sustained interrogation of the conventions of pictorial illusionism. (Cottington, 1998, p 133)

Unlike many of their artistic predecessors generating aesthetic innovations in the 19th century, Braque, and especially Picasso, also differentiated themselves in the 1900s by not seeking legitimacy and acclaim through the public display of their works in the Academy-organized Salons to reach audiences and collectors. Particularly after the production of *Les Demoiselles d'Avignon* in 1907, Picasso stayed aloof from these established means of gaining attention and support, relying instead on the commitment of private galleries, notably that of Kahnweiler (Cottington, 2004, pp 14–8). Despite this seclusion from the public French art world, Cubism:

> was an idea of such power and flexibility that it spread across Europe and the Americas with astonishing speed, thanks in part to (avant-garde)

> artists from across the globe who became disciples of the new art. Many of these had in fact seen relatively little of the strange images made by Picasso and Braque. (Cox, 2000, p 4; see also Chapter 5)

These four kinds of innovations need to be distinguished from fads and fashions, or short-term enthusiasms, which are likely to develop in fields with low mutual dependence if authority over standards and resources is widely distributed. While developing innovations are processes that bring about lasting change, fads and fashions are transient changes of frames of reference that are reversible and disappear when new frames of reference replace the current ones (following Simmel, 1957 [1904]; Blumer, 1969; Gronow, 2009). They differ from hypes that can be understood as an exaggerated response by practitioners to findings that promise new aesthetic/epistemic opportunities (applying ideas from the hype cycle model of technical innovation but not the full model, Fenn and Raskino, 2008; Dedehayir and Steinert, 2016). They thus temporarily exaggerate a process that nevertheless leads to lasting change (Gläser et al, 2026).

Innovation processes and complementarity in the socio-economic environment

Distinguishing between innovations in the modern arts and sciences in terms of the nature of the discontinuous changes they generate enables us to compare the development of different kinds of changes in different fields and contexts. In particular, this approach focuses attention on the processes by which novelties are developed as distinct types of innovations and create a variety of changes in artistic and scientific practices and purposes in contrasting circumstances. This in turn highlights the contingent nature of innovation processes and their dependence on complementary developments (Rosenberg, 1975), both within practitioner communities and in their wider organizational and socio-economic contexts as Rosenberg and others have emphasized.

As the cases of BEC, Ethnomethodology and Metafiction illustrate, these contingencies not only affect how particular novelties become distinct kinds of innovations in individual fields, but can also lead to changes in their nature and impact on practitioners' novelty production practices and contributions. The production of BEC, for instance, was initially a significant experimental achievement in atomic, molecular and optical physics that stimulated a number of attempts to generate similar phenomena with a variety of atoms in different European countries. As techniques and skills became more standardized and transferable between laboratories and researchers, though, BEC production practices for some, but not all, atoms developed into established platforms for investigating problems and questions in quantum mechanics and so constituted a major component of a complementary innovation in low-temperature physics. What was a significant process innovation in the 1990s has thus led to a considerable shift in questions asked and pursued in broader areas of physics.

The development of Ethnomethodology and Metafiction novelties as innovations similarly changed their radically disruptive nature and impact in the course of becoming established, from initially pursuing a largely displacement strategy to effecting more complementary changes in practices and dominant problems. Both sought to change the established conventions and standards governing the production of recognized contributions to collective purposes by attacking the dominant assumptions and procedures in North American sociology and Anglophone fiction, respectively, together with producing disruptive contributions exemplifying their approaches. After initial responses of both aggressive counter criticisms and indifference to their claims and outputs, changes to the organization of both fields and their socio-economic contexts provided complementary inputs for their eventual acceptance as legitimate additions to established practices and standards, albeit perhaps more widely in the case of Metafiction than for Ethnomethodology (Klinkowitz, 2017; Hammersley, 2018).

In other examples of significant innovations, complementary shifts in aesthetic and epistemic practices and conventions, in addition to organizational and socio-economic changes were often important in ensuring that some novelties did generate substantial changes, as suggested by Rosenberg (1975) in economics. In the case of the development and establishment of Radio Astronomy as a significant process innovation within astronomy, for instance, Martin (1978) has suggested that the parallel debate in cosmology over the structure and evolution of the universe helped experimentalists in Cambridge to gain financial support and broader scientific interest for their mapping of radio sources beyond our galaxy. Similarly, technological developments can provide complementary support for significant changes in scientific fields, such as the transformation of business finance and investment teaching and research into Financial Economics. Here, the production of computers capable of processing large amounts of information in the 1950s and 1960s, coupled with the reorganization of financial markets facilitated the rapid growth of research on stock-market price movements and the widespread acceptance of the efficient market hypothesis and capital asset pricing model (Whitley, 1986b; Chapter 5).

In the arts, complementary changes in production processes and aesthetic preferences have often been combined with major shifts in the nature of authoritative intermediaries and audiences to generate discontinuous and disruptive changes in novelty-production procedures and priorities. The disruption of dominant conventions and values and eventual acceptance of Impressionism as an aesthetic innovation during the 19th century, for instance, was considerably aided by the emergence of the 'dealer-critic' system for supporting, distributing and evaluating new works in conjunction with the growth of the middle-class market for genre paintings as well as being supported by the expansion of commercial production and distribution of pigments in tubes, the development of the 'French box' means of transporting equipment for easel painting and the weakening of established aesthetic hierarchies in state-supported Salons and academies (White and White, 1965; Cottington, 2022; Chapter 5). Similar changes in the composition and size of cinema audiences and in the distribution of films in combination with postwar shifts in state policies and technical advances helped to complement and support the aesthetic interests and concerns of the *Nouvelle Vague* critics and directors in the French film industry in the late 1950s and early 1960s (Rachlin, 1993; Marie, 2003; Parkinson, 2012; Chapter 5).

The four kinds of innovation processes established in this chapter have received uneven attention in studies of change in the arts and sciences. Most empirical studies focus on the more consequential changes, that is, on complementary and displacement innovations. Although the focus of the literature has been on these two innovations, and often on the innovation-triggering events rather than the processes of innovation development following them, there are a number of rich empirical studies of process, complementary and displacement innovations that enable a comparative analysis of innovation processes in different artistic and scientific fields to be conducted. How such innovations are developed depends on their specific societal contexts, which therefore need to be systematically included in the study of innovations. We consider the structure of these contexts in the next chapter.

4

The Social Contexts of Innovation Development in the Modern Arts and Sciences

The different kinds of innovation and change of artists' and scientists' practices and purposes discussed in the previous chapter are developed and implemented by practitioners in the specialized social contexts established by the institutionalization of the modern arts and sciences as novelty-driven fields of cultural production summarized in Chapter 2. These contexts are the practitioner community in which novelty-driven production takes place and the art and science worlds in which the recognition of novelties and the allocation of resources for novelty production occur. In this chapter, we develop a comparative framework of art worlds and science worlds that captures important variations in the nature of these social contexts that directly affect (a) practitioners' decisions to develop an innovation and (b) their ability to implement these decisions.

In the first section of this chapter, we consider the situation of practitioners who are confronted with the opportunity to develop an innovation and outline the major factors that influence the development of innovations: the nature of productive relationships in practitioner communities and the distribution of authority in art worlds and science worlds. These two factors are discussed in the second and third sections. Understanding practitioner communities as producers of contributions to shared aesthetic/epistemic problems directs attention to the different kinds of productive relationships that they develop in producing novelties. Contrasting dominant patterns of productive relationships leads to two ideal types of practitioner communities that affect different kinds of innovations.

Variations in prevalent patterns of productive relationships in practitioner communities and in the distribution of authority in art worlds and science worlds influence how creative practitioners respond to each other's novelties, and how these responses result in the development of different kinds of innovations. In the third section, we suggest how variations in the distribution of authority in art worlds and science worlds can be expected to affect the organization of different artistic and scientific fields and innovation development. Combining types of practitioner communities with types of art worlds and science worlds in the fourth section enables us to analyse the development of different kinds of innovations in different socio-economic contexts.

Influences on the development of innovations

The immediate social context for the development of innovations is the community of creative practitioners dealing with shared aesthetic or epistemic problems. Producers of novel contributions consider the purposes and standards of their communities when deciding to change their practices for the development of an innovation. Their ability to implement such decisions depends in particular on: (a) their autonomous control of resources over the period in which the change of practices occurs and (b) the expectation that the innovation will be recognized by the community as a significant opportunity for dealing with its major problems. Both of these conditions depend in turn on the interactions of practitioner communities with the wider societal contexts in which they are embedded, including a variety of non-practitioner audiences and intermediaries that perform different roles in recognizing, interpreting and distributing novelties in art worlds and science worlds.

Creative practitioner communities in the modern arts and sciences belong to a distinct subtype of community in which their members produce responses to each other's novelty production in their creation of new outputs. From the definition of novelty-driven production in Chapter 2, it follows that a creative product's novelty and relevance are established *vis-à-vis* those of others addressing the same set of aesthetic/epistemic problems. Members of practitioner communities are aware that they produce new contributions dealing with shared questions and adhere to common standards of relevance and quality of novelties. To do so, they respond to each other's contributions, that is, they engage in productive relationships with at least some other community members. These productive relationships range from acknowledging each other's novelties as being relevant to collective purposes to using them as material and transforming them in their own production of novelties to cooperatively producing novelties. They affect creative practitioners' decisions on what to produce and how to produce it.

Productive relationships between practitioners affect their decisions to develop innovations in two ways. First, they influence whether practitioners perceive a colleague's novelty to be relevant to their own novelty production.

Community members may consider an innovation developed by others not to be relevant to their own work because it addresses different aesthetic or epistemic problems or because it utilizes a different approach. Second, the decision to develop a relevant innovation depends on the costs and benefits involved, including the costs and benefits of changing productive relationships. Developing innovations means changing both practices of novelty production and the nature of productive relationships with different groups of colleagues. These changes, their costs and the opportunities for the further production of novelties are necessarily involved when making a decision to develop an innovation.

An important potential benefit of developing an innovation concerns its expected usefulness for the future production of relevant novelties. Where changes in practices and purposes promise to provide better opportunities for relevant novelty production and thus for an extension of productive relationships, they are likely to enhance the appeal of working on the innovation. Productive relationships can be extended and improved through the production of novelties that are either more relevant to the aesthetic or epistemic interests of the same producers and audiences or become relevant to a larger number of practitioners and audiences.

In addition to the initial costs and benefits of changing productive relationships affecting innovation development, the change of practices involved often creates further risks and delays. Overall, three main risks can be distinguished. First, the legitimacy of an innovation may be unclear in the initial phase of its development because it may not yet meet established standards of quality and relevance. Second, producing novelties that develop the innovation may require more time than expected due to necessary investments in learning new skills and uncertainties in the production process. Third, available resources may be insufficient to achieve the change of practices and purposes. These risks may considerably delay the production of novelties that develop the innovation or may even prevent it altogether if they cannot be readily accommodated by practitioners.

Owing to these risks, the development of innovations in novelty-driven fields requires a minimum amount of **protected space** for practitioners. The notion of protected space has been used in innovation research to describe a social space in which the development of technological innovations is protected from unfavourable conditions (Rip, 1995, 2011). The need for protected space exists not only at the collective level of practitioners jointly developing an innovation but also at the individual level of creative practitioners who produce novelties and/or develop an innovation. Protected space is here understood as *the period of time for which creative practitioners have control over the use of particular amounts of human and material resources to pursue particular problems without suffering severe reputational and career consequences* (Whitley and Gläser, 2014, p 8; Whitley et al, 2018). It has two dimensions: first, the length of time, or duration, for which practitioners are able to exercise this control and second, the extent of independent control of resources, including the proportion of working time available for researchers to devote to particular problems.[1]

The three risks of unclear legitimacy, running out of time and inadequate resources are inherent to any production of novelty, but developing innovations usually requires particularly high levels of protected space due to the change of practices and priorities involved and the resulting additional uncertainties. In their decisions to produce particular novelties and to develop innovations, creative practitioners take into account the level of available protected space relative to the amount that they expect to be necessary for the task (see Fujimura, 1987, on the 'do-ability' of research problems). They may build protected space by acquiring resources and negotiating autonomous control over them for the period they deem necessary for the change of practices, or they may utilize protected space they can access or which is created for them by others.

Opportunities to build and utilize protected space by gaining control over human and material resources depend greatly on the distribution of authority between practitioners, intermediaries and audiences in art worlds and science worlds. Authority is here understood as legitimate influence (Rosenau, 2004, p 32) and is exercised over aesthetic/epistemic practices, purposes and standards governing the allocation of resources for producing, distributing and evaluating outputs. Much of this authority is exercised through the procedures and criteria for distributing novelties to audiences and allocating resources to individual and collective practitioners.

Productive relationships in practitioner communities

For innovations to be developed, they must be expected to provide new opportunities for producing novelties contributing to collective purpose in addition to being recognized as legitimate contributions. In particular, they must be considered likely to enable contributions dealing with legitimate aesthetic/epistemic problems which outweigh the costs of changing practices and purposes, including the efforts involved in convincing practitioner communities of the significance of one's own approach for collective aesthetic/epistemic goals (for the sciences, see Whitley, 2000, pp 87–8).

The costs and benefits of developing an innovation in a practitioner community depend on the prevalent ways in which practitioners respond to and make use of each other's work in their own novelty production. There are two distinct ways in which practitioners can take account of, and use, others' novelties in developing their own production: *positional* use and *absorptive* use. *Positional use* of colleagues' works accompanies practitioners' recognition of each other as members of the same community in which relevant novelties have to contribute to the solution of shared problems and questions. Practitioners relate their novelties to those of others, either convergently or divergently, in order to demonstrate how their contribution addresses their community's shared aesthetic or epistemic concerns. This positional use of others' novelties is ubiquitous in the modern arts and sciences as they became institutionalized as novelty-driven fields of cultural production.[2]

Such explicit positioning of one's work often emphasizes the divergence between its focus and approach from those of others. For example, in the introduction to her book *Narcissistic Narratives: The Metafictional Paradox*, Linda Hutcheon makes it clear that she is:

> avoiding the 'postmodern debate' in in this particular kind of study. Most discussions of 'postmodernism' are concerned primarily with the psychological, philosophical, ideological or social causes of the flourishing self-consciousness of our culture. This book, however, makes no pretence to contributing to the interesting controversy that has engaged Robert Alter, James Sloan Allen, Gerald Graff, and Ihab Hassan, among others. Instead, it merely accepts that something has happened and that it has its roots deep in our culture, most strongly in the Romantic and modernist period. (Hutcheon, 1980, p 3)

Similar to this positioning of a novelty against others in literary criticism, advocates of what became known as total serialism in art music in the aftermath of the Second World War emphasized their opposition to, and remoteness from, neoclassical conventions and practices. Some, such as Boulez, additionally attacked composers who had been considered radically deviant earlier in the 20th century, including Stravinsky and even Schoenberg, for being insufficiently purist and exhibiting romantic tendencies (Ross, 2012, pp 393–6).

The more common practice of convergently relating novelties to those of others often includes the use of others' outputs to demonstrate how one's work is relevant to established approaches, usually in support of the approach adopted. While not typically directly used in the production of one's own novelties, such discussions of particular works serve to identify how they fit into previously established frameworks and problem formulations, clarifying their relevance to commonly accepted questions and the ways in which they improve on current understandings. For example, in his 'Notes of a Painter' Henri Matisse suggested that he is like Cézanne in developing and following a clear conception of a new work that avoids confusion of expression (Bock, 1981, p 196).

Similarly, Patricia Waugh has emphasized that metafictional writers continued to rely on many of the conventions of 19th-century literary realism when conducting fictional experiments that parodied them in order to undermine the authority of the omniscient author, of the closure of the final ending and of the definitive interpretation.

> Metafiction explicitly lays bare the conventions of realism; it does not ignore or abandon them. Very often realistic conventions supply the 'control' in metafictional texts, the norm or background against which experimental strategies can foreground themselves. … Metafiction, then, does not abandon 'the real world' for the narcissistic pleasures of the imagination. What it does is to re-examine the conventions of realism in order to discover – through its own self-reflection – a fictional form that is culturally relevant and comprehensible to contemporary readers. In showing us how literary fiction creates its imaginary worlds, metafiction helps us to understand how the reality we live day by day is similarly constructed, similarly written. (Waugh, 1984, p 18)

An example of convergent positioning in the sciences is the response by some researchers to the development of Evolutionary-Developmental Biology (Evo-devo). Evo-devo is a highly heterogeneous interdisciplinary field that explores links between the evolution of a species (the subject matter of evolutionary biology) and the embryonic development of its individuals (the subject matter of developmental biology). Researchers can commit to contributing to this field in a variety of ways that incur different costs and risks (Chapter 5). The easiest way to engage in Evo-devo research was to continue an Evolutionary Biology or Developmental Biology research line including the presentation of findings to the traditional audiences but additionally to position findings in the Evo-devo conceptual framework. This positioning of experimental findings in both traditional and Evo-devo research contexts did not require any changes in experimental strategies or designs (Laudel et al, 2014a, pp 243–4).

The second way in which practitioners use others' novelties, *absorptive use*, occurs when artists and scientists directly build upon, and incorporate, others' novelties into their own production processes. Absorptive use in the arts frequently occurs when techniques, motifs or designs developed by other artists are made into parts of a new artwork, as in the numerous 'borrowings' from their predecessors' works by art music composers in the 19th and 20th centuries (Morgan, 1978; Sallis, 2018). In the case of Charles Ives' compositions, Peter Burkholder has identified no fewer than 14 distinct types of integrating others' works into his own compositions, including:

(1) ***Modelling*** a work or section on an existing piece, assuming its structure, incorporating part of its melodic material, imitating its form or procedures, or using it as a model in some other way …;
(2) ***Variations*** on a given tune …;

(3) ***Paraphrasing*** an existing tune to form a new melody, theme, or motive …;
(4) ***Arranging*** a work for a new medium …;
(5) ***Setting*** an existing tune with a new accompaniment …;
(6) ***Cantus firmus***, presenting a given tune in long notes against a more quickly moving texture …;
(7) ***Medley***, stating two or more existing tunes, relatively complete, one after another in a single movement …;

…

(12) ***Collage***, in which a swirl of quoted and paraphrased tunes is added to a musical structure based on modelling, paraphrase, cumulative setting, or a narrative program … . (Burkholder, 1994, p 854)

In painting, Michael Baxandall has highlighted Picasso's use of Cézanne's pictures:

> But again, and very obviously, Picasso also went to Cézanne's pictures as an actual resource, somewhere he could find means to an end, varied tools for solving problems. The matter of Cezannian passage – of representing a relation between two separate planes by registering them as one continuous superplane – I have already mentioned, but there are other things too Picasso is considered to have adapted from Cézanne: for instance, high and sometimes shifting view-points that flatten out on the picture-plane arrays of objects phenomenally receding in depth. (Baxandall, 1985, p 61)

Absorptive use also occurs in the social sciences and humanities. In his analysis of the *Nouvelle Vague* in the French film industry, Rachlin describes a group of directors as 'the leaders of a French film subculture' using Hebdige's definition of 'subculture':

> In using the term 'subculture' here and throughout this paper, I follow Hebdige's work in which subculture is seen as a 'refusal' or 'self-imposed exile' from the mainstream (Hebdige, 1979: 2–3). I feel it is an appropriate term here, because of all the disparate elements tying together the men who would make the Nouvelle Vague, none was, as I will show, more significant than their utter disgust with the products of mainstream French film. (Rachlin, 1993, p 432)

In the physical and biological sciences, absorptive use is common because in many cases relevant novelties cannot be produced without it. For example, the experimental realization of Bose-Einstein Condensation (BEC, a specific state of matter occurring at very low temperatures) used as input, among other contributions, both the prediction by Einstein that such a physical state of atoms exists and knowledge about previously developed cooling techniques:

> In brief, the conditions for BEC in alkali gases are reached by combining two cooling methods. Laser cooling is used to precool the gas. The principle of laser cooling is that scattered photons are on average blue-shifted with respect to the incident laser beam. As a result, the scattered light carries away more energy than has been absorbed by the atoms, resulting in net cooling. Blue shifts are caused by Doppler shifts or Stark shifts. The different laser cooling schemes are described in the 1997 Nobel Lectures in Physics. After the precooling, the atoms are cold enough to be confined in a magnetic trap. Wall-free confinement is necessary, otherwise the atoms would stick to the surface of the container. It is noteworthy that similar magnetic confinement is also used for plasmas, which are too hot for any material container. After magnetically trapping the atoms, forced evaporative cooling is applied as the second cooling stage. (Ketterle, 2002, p 738)

These inputs informed the design of experiments and thus their outcomes (Laudel et al, 2014b, pp 211–3). Had any of the inputs been different, different experiments would have been developed and yielded different results. Thus, the novelties absorbed by those who produced BEC became an inextricable part of these condensates. Similarly, developing research on the genetic causes of cancer is only possible if specific molecular-biological methods are used. Fujimura (1988; 1996, pp 68–115) has shown how the standardization of molecular-biological methods made it possible for cancer researchers to incorporate them in their research and to address new research questions about the role of genetic mutations in causing cancer.

These two kinds of productive relationships, positional and absorptive, extend throughout practitioner communities as both direct and mediated productive relationships. Practitioners maintain direct productive relationships by positioning their novelties against some of the novelties produced by other community members and by using the novelties of some other community members as input for their work. These productive relationships form networks that link some members of the community, and the overlapping networks of productive relationships extend throughout the community. In the sciences, these relationships are partly reflected in networks of citations, which are used by researchers to make their productive relationships explicit. However, these networks are always incomplete because some productive relationships will not be acknowledged for a variety of reasons. For example, an analysis of journal articles from economics and management containing the concept ‘bounded rationality’ showed that only about 27 per cent also contained explicit references to works by Herbert Simon, who introduced the concept (McCain, 2015, p 2191).

Creative practitioners can maintain such direct productive relationships only with some of their fellow community members. It is impossible for them to position themselves against all other community members because their work is too diverse. Even positioning one's work *vis-à-vis* a community's mainstream creates productive relationships with a high proportion of community members but not all of them. The number of fellow community members whose contributions can be incorporated through absorptive use is likely to be much smaller.

In addition to these direct productive relationships with a limited reach, community members are also interlinked by indirect productive relationships which are mediated by shared aesthetic/epistemic problems and shared resources like data, methods and techniques, conceptual approaches, theories, objects and materials. If practitioners address the same aesthetic/epistemic problems, they affect each other's novelty production to the extent that their contributions change the understanding of that problem. Similarly, contributions to shared aesthetic/epistemic resources change them and may affect the work of practitioners who use them. Productive relationships mediated by shared aesthetic/ epistemic resources can be both positional if practitioners position themselves *vis-à-vis* resources and absorptive if they use them.

The most obvious example of such indirect productive relationships is the use of theories, conceptual perspectives and methods in the sciences, social sciences and humanities. While sometimes other researchers are mentioned as protagonists of theories, approaches, or methods, they are rarely the only practitioners who contributed to their development. Epistemic resources like Quantum theory, feminism or statistical methods are jointly developed by researchers, and contributions to that development create indirect productive relationships with other community members.

Similar to this sharing of epistemic resources in the sciences, albeit usually less explicitly, artists share aesthetic resources when addressing shared aesthetic problems. For example, until Cubism was developed, the assumption that paintings must present an illusion of reality at a particular moment in time guided the work of most painters, who also applied traditional techniques like using light and shade to create the illusion of volume, foreshortening objects to imply depth, and portraying subjects from a single viewpoint and perspective (see Chapter 5). Similarly, art music composition shared a 'common practice tonality' of pitch and rhythmic conventions until it was challenged by atonal composers at the end of the 19th century (discussed later).

These direct and indirect productive relationships integrate practitioner communities by linking each member directly to some other members, by forming overlapping networks of such links that extend across the community and by linking community members who share aesthetic/epistemic resources. The integration of a community itself takes different forms depending on how networks of direct productive relationships are linked and how aesthetic/epistemic resources are shared between community members.

Two ideal-typical patterns of productive relationships can be distinguished depending on the scope of sharing aesthetic/ epistemic problems and resources. First, **Integrated communities**, in which most practitioners are affected by, and need to respond to, the novelty production of other community members because aesthetic/epistemic problems, approaches and objects are widely shared across the community. Second, **Polycentric communities**, in which strong productive relationships develop between groups of practitioners who share some of the aesthetic/epistemic problems addressed by their community, some of the approaches used in their community, some of the objects community members work with, or any combination thereof.[3] Integrated and Polycentric communities are ideal types that represent the poles of a spectrum. The stronger and more extensive a community's shared background assumptions are, the more integrated it is. These two types of communities are likely to differ in their impact on the development of different kinds of innovations in practitioner communities.

Considering *Integrated* communities first, networks of direct and indirect absorptive relationships dominate and extend throughout the practitioner community because the community's shared aesthetic/epistemic problems are closely interconnected and aesthetic/epistemic resources are shared by most community members. Practitioners must use some of each other's novelties because relevant contributions cannot be produced without them. While the relevant state of the art and absorptive relationships might not be the same for all practitioners, the strong shared background of aesthetic/epistemic resources results in community-wide criteria for the assessment of relevance and quality of contributions. The widespread sharing of aesthetic/epistemic resources here produces a high level of mutual dependence between practitioners.

In the case of atomic and molecular optics, researchers addressing problems of BEC and their colleagues working on atomic spectroscopy would rely on the same theoretical background and would use the same methods for cooling atoms, which means that some standards of relevance and quality would apply to them regardless of the problem they address and the approach they are using. The findings of researchers working on one problem are likely to become immediately relevant to their colleagues because they change shared theories or methods, or because they contribute knowledge about empirical objects many community members work with.

This mutual dependence of practitioners in Integrated communities is exemplified by physicists' comments on the news that the OPERA experiment at CERN detected neutrinos moving faster than light:

> If the OPERA results are confirmed they will call for a *complete reconsideration of the basic principles of particle physics*. In a Lorentz-invariant theory, propagation of neutrinos faster than light would imply violation of causality (i.e. the CNGS beam becomes GSCN when seen from a sufficiently boosted reference frame). It is extremely hard to envisage a consistent theory having this property. (Giudice et al, 2012, p 3, our emphasis)

The integration of the particle physics community is such that a single experimental result like the observation of superluminal neutrinos would require a reconsideration of fundamental theoretical principles which in turn affect the whole particle community. While this integration through a shared theory that makes results contributing to it being absorbed by all practitioners using the theory may constitute the strongest case of mutual dependence in Integrated communities, similar relationships also exist in many scientific communities that do not share unified theories to the same extent. For example, if most community members work with the same empirical objects, their empirical results need to be consistent with the body of existing empirical knowledge shared by their community.

Integrated communities can be found in many of the sciences, particularly in those fields where most theories and methods are shared by all community members. Conversely, many practitioner communities in the arts and sciences are less integrated in that productive relationships are mainly developed between members of different groups that use the same or closely similar conceptual or methodological approaches or use the same empirical objects. For example, in sociology, institutionalists address the problem of social order differently from ethnomethodologists, and in history medieval historians have other concerns than do early modern ones. In painting, Picasso interpreted the problem of representing three-dimensional situations and stories in a two-dimensional surface in different ways to the established academic tradition (Baxandall, 1985, pp 44–5). Thus, where creative practitioners are committed to different approaches in dealing with constitutive aesthetic/epistemic problems and/or focus on different problem areas, they are likely to establish different kinds of productive relationships with colleagues sharing these commitments from those that they establish with colleagues committed to different problems and/or approaches.

In such *Polycentric* communities, practitioners are committed to dealing with particular aesthetic/epistemic problems and approaches to solving them. Practitioners still need to position themselves not only *vis-à-vis* those with whom they share their approach but also in relation to other approaches and to the shared community-wide background. This arrangement of productive relationship is common in the social sciences, the humanities and the arts.

In the field of late 19th- and early 20th-century art music composition, for instance, advanced composers of the period developed a variety of responses to the problem of 'the progressive deterioration of the pitch and rhythmic conventions of so-called "common practice tonality" (when) the basic structural functions of the tonal system became more and more equivalent tending to level out all musical statements and thus render them increasingly interchangeable' (Morgan, 1978, pp 73–4). While some of these composers, such as Schoenberg and his students, rejected all melodic repetition, variations and other forms of musical coherence and comprehensibility for a brief period (Whittall, 2009), others retained established conventions while developing new conceptions of the nature and limits of serious musical language, including Charles Ives and Gustav Mahler (Morgan, 1978; Schorske, 1983).

As Robert Morgan has pointed out:

> Yet both composers often cling to an historically regressive stage of the musical language of common practice tonality, a stage long since passed over by the currents of nineteenth century evolutionary chromaticism. The chromatic saturation of the tonal field, so consistently evident in other advanced composers of the time, is often absent in Ives and Mahler, even in their later compositions … In both Ives and Mahler tonality in the strict, functional sense remains an active force. (Morgan, 1978, p 73)

Similarly, there seems general agreement that the study of History in the postwar USA is more integrated around common standards governing research practices than are contributions in other fields in the Humanities despite the diversity of practitioner groups concerned with different questions and approaches (Lamont, 2009, pp 81–4), As reported by Lamont (2009, p 80), the panellists deciding on the allocation of research grants and fellowships in History in her comparative study of the humanities and social sciences shared a broad agreement on what constituted good historical craftsmanship and careful archival work. While in no way implying the field's unification around a common theory, this enabled them to reach joint decisions on the quality of applications relatively quickly.

Integrated and polycentric patterns of productive relationships in practitioner communities can be expected to affect the perception of costs and benefits of developing different types of innovations. The strong mutual dependence between practitioners in Integrated communities, for instance, makes it necessary for them to consider colleagues' novelties by either rejecting them or by further developing them. While often either is possible due to the broad range of opportunities for novelty production available to practitioners, a specific decision is usually required.

For example, the decision for or against developing an innovation based on the discovery of superluminal neutrinos was necessary because the theoretical foundations of much further work in that area were concerned. Delaying the decision on accepting or ignoring the finding would have meant delaying the decision on the correctness of the theoretical foundations of the field, which would severely constrain the future production of novelties. Similarly, the discovery of 'cold fusion' required such a decision because the findings were at odds with the theoretical foundations of physical chemistry (see Chapter 5).

The predominance of absorptive productive relationships in Integrated practitioner communities makes the development of *displacement* innovations contingent on the innovations being widely understood to be superior to existing approaches for producing novelties. When novelties fail to promise superior opportunities for novelty production, they are not developed at all or abandoned quickly, as happened in the case of superluminal neutrinos and cold fusion. However, if they are so regarded, they have better chances of becoming displacement innovations than in Polycentric communities because they affect approaches that are used throughout the community and thus provide opportunities to produce superior novelties for many practitioners. The same is true for *process* innovations, which need to promise superior novelty production soon but will be developed rapidly if they do. *Complementary* innovations can also easily be developed in Integrated practitioner communities if they promise superior opportunities for novelty production and are consistent with the knowledge shared by the community.

In Polycentric communities, the mutual dependence within centres – practitioner groups committed to a particular set of aesthetic/epistemic problems and a common approach to them – is stronger than their connections with practitioners belonging to different centres. This implies that the conditions for developing *process* innovations within each centre are usually similar to those in Integrated communities. Such innovations are likely to be developed when they provide superior opportunities to produce novelties within the adopted framework.

Complementary innovations that lead to new centres being established besides existing ones are likely to have high chances of success provided that sufficient resources are available and new practitioners and audiences can be recruited for the innovation. For example, the philosophy of information became institutionalized in the second half of the 20th century in response to the development of computing (Floridi, 2002). The philosophy of information can be defined as 'the philosophical field concerned with (a) the critical investigation of the conceptual nature and basic principles of information, including its dynamics, utilization and sciences, and (b) the elaboration and application of information-theoretic and computational methodologies to philosophical problems' (Floridi, 2002, p 137). Other examples of the development of new subfields of the social sciences and humanities that develop new topics include 'climate history' (Duffy, 2024) and the exploration of new approaches to sociological questions like the 'sociology of valuation' (Lamont, 2012).

In contrast, complementary innovations are less likely to be developed within existing centres when they are seen to undermine the presumptions and practices of an existing approach by proposing alternative questions and approaches. Such innovations are more likely to be separated and to lead to a new understanding of major problems and appropriate approaches. The introduction of quantitative, internationally comparative, student assessments, such as PISA studies, in German educational science, for instance, was incompatible with the traditional philosophical approach to *Bildung* as self-development, which so far had dominated German educational research. Researchers advancing the new quantitative approach of measuring student competencies split from those adhering to a more holistic approach to developing students' personalities, which led to a separate institutionalization of German quantitative educational science (Aljets, 2014; Gläser et al, 2014; Zapp and Powell, 2016). In contrast, complementary innovations that enlarge the range of questions and problems without challenging the legitimacy of existing aesthetic/epistemic problems and approaches are more likely to be developed.

Displacement innovations in Polycentric practitioner communities are possible only to the extent that they replace shared background assumptions that can integrate disparate approaches. For example, in the polycentric field of research on cancer at the end of the 1970s, the shared assumption that cancer is caused by a particular 'factor' in the cell generating cancer was displaced by the shared 'Oncogene Approach', that is the assumption that cancer is caused by some normal cellular genes being turned into cancer genes (see Chapter 5). This new collective background assumption led to a stronger integration of research on genetic, viral and environmental causes of cancer. If intended displacement innovations fail to change the community's shared background, they are likely to be turned into complementary innovations that add a new centre to the existing ones. This happened to Ethnomethodology, which challenged the foundations of sociology and was originally developed to replace traditional sociology but ended up as one of the many competing approaches that coexist in the discipline (Mullins, 1973).

The extent to which practitioner preferences for developing particular kinds of innovations can be realized and the speed with which this is possible depends not only on the nature of productive relationships but is strongly affected by the authority structures of art worlds and science worlds. In particular, the authority of audiences and intermediaries over the recognition of contributions and their influence on the allocation of resources for the development of innovations can make the development of different types of innovation more or less likely. We therefore turn now to consider patterns of authority distribution in different art worlds and science worlds.

The distribution of authority in art worlds and science worlds

The institutionalization of the arts and sciences as novelty-driven fields depended on practitioner communities gaining substantial collective authority over the standards governing: (a) the recognition of novelties as relevant contributions for future novelty production, (b) the competence of practitioners and (c) the allocation of economic resources.

Such practitioner collective authority varies in two major ways that affect practitioners' ability to construct and utilize protected space and thus the development of innovations. First, in its degree of aesthetic/epistemic unity or diversity of goals, questions and approaches adopted by competent practitioners and, second, in the extent to which it is shared with incumbents of other roles in art worlds and science worlds, particularly interpretative and commercial intermediaries, and with different kinds of external audiences. We first discuss variations in the level of unity or diversity of aesthetic/epistemic authority in creative practitioner communities before considering variations in the authority of different intermediaries and audiences over practices and purposes in art worlds and science worlds.

The distribution of authority over the collective standards governing the nature of competent and significant contributions to a field's goals varies considerably between practitioner communities, as well as over time. In particular, it differs between Integrated and Polycentric communities in the extent to which it is divided between different approaches and conceptions of the community's purposes and appropriate practices. In Integrated practitioner communities, most standards governing the recognition of novelties as relevant and useful contributions apply throughout the community. These communities thus are not only integrated with regard to their productive relationships but also possess an integrated authority structure governing novelty production practices and assessments across different problem areas and approaches. An integrative authority structure exists when regardless of which community members exercise authority, they do so in the same way, that is, by making similar decisions based on the same standards and decision criteria.

Conversely, Polycentric practitioner communities are divided between groups addressing different aesthetic/epistemic problems and/or using different approaches, thereby extending the range of the community's novelty production. The aesthetic/epistemic authority over standards, priorities and the provision of resources is distributed between such groups, and the authority structure of these communities is thus polycentric. A well-known example is the rivalry between quantitative and qualitative social research and the resulting fight of qualitative researchers for adequate standards of quality (for example, Denzin, 2009).

In German educational science, the rise of empirical (quantitative) education research led to a split of the German Association of Educational Research and to a separate institutionalization of quantitative empirical education research at several German universities (Gläser et al, 2014). In contrast, the focus of historians on different historical periods and regions and their use of competing approaches to the reconstruction and interpretation of historical processes did not affect the shared background of the field but authority over the assessment of the relevance of a contribution was distributed between groups of practitioners adhering to the same approach or answering questions about the same historical period.

In the arts, the musical standards and purposes developed by art music composers and performers in 1960s and 1970s New York City differed quite sharply between the 'uptown' and 'downtown' approaches to new music production and performance according to Samuel Gilmore (1987, 1988). While the former tended to pursue serialist and neo-tonalist approaches in regular but not frequent concerts and benefitted from academic institutional support, the latter pursued more experimental goals and approaches, such as minimalism, drone and aleatory or 'chance' music, with irregular and inexpensive performances by friends and colleagues.

> (The) Uptown (compositional sub-world) offers more concert resources than Downtown, but to avail themselves of this financial and performance support, composers must become academics, since it is through the university that Uptown composers gain access to jobs, concert halls, performers and other concert resources. Downtown composers, on the other hand, primarily generate resources through non-musical sinecures, performance space in their own lofts, and interpersonal arrangements … Uptown composers use compositional practices that constrain instrumentation, notation and compositional theories, while Downtown practices allow innovation in all these areas … Uptown notation and compositional theories incorporate less radical change than Downtown and seek to maintain a musical tradition … Downtown notation is graphic and open to interpretation, when it is used at all. (Gilmore, 1988, pp 213–4)

In this case, authority over access to performers, training, performance venues and other resources was heavily weighted to Uptown 'twelve tone' composers and their aesthetic choices. 'In the late sixties and early seventies, twelve tone composers were reaching the height of their influence. By some accounts, they effectively took control of university composition departments across the country … young composers with tonal yearnings found little happiness in academia' (Ross, 2012, p 531).

In principle, where aesthetic/epistemic authority is distributed between a number of different approaches and concerns, we would expect practitioners to find it easier to gain initial tolerance for unusual and challenging contributions than in more unified fields. As some variants of American minimalism were modified into less disruptive musical styles after the 1960s and 1970s, for instance, they eventually became more widely recognized as art music and gained larger audiences (Bernard, 2003; Levaux, 2020, pp 165–81). In contrast, in the case of aleatory (chance or random) literary works many of those pursuing comprehensively experimental strategies that rejected most established conventions struggled to gain colleagues' commitment to develop their outputs for their own novelty production because there was no shared background that could be used to relate novelties to each other (Waugh, 1984, p 12; Chapter 5).

Considering next variations in how practitioner communities share their authority over the recognition of novelties as legitimate contributions to the epistemic/aesthetic purposes of their field with incumbents of other roles in art worlds and science worlds, these differ primarily in the means used to exercise authority and in the extent of audience and intermediary involvement. They include: (a) the standards of relevance and quality that develop in practitioner communities and can be influenced and applied by knowledgeable intermediaries and audiences, (b) the reputation of creative practitioners and intermediaries, and (c) the material resources that can be invested in the further production of novelties, which usually include contributions to the subsistence of creative practitioners. Access to these means of exercising authority varies both among and between the incumbents of the three major types of roles in art worlds and science worlds as well as over time. Consequently, the use of these means to build or grant protected space for the development of innovations differs between art and science worlds depending on the interests and authority of incumbents of different intermediary and audience roles.

In art worlds, authoritative roles of intermediaries overlap because interpretive, distributive and constructive intermediaries all affect the flow of funding to the production of novelties, be it by making novelties accessible to paying audiences, by guiding audiences' attention and assessment or by channelling state funding. They thus can exercise some, and under some circumstances considerable, authority over which novelties can be produced at all, which creative products find an audience, how they are assessed, and how these assessments by audiences affect the further production of novelties.

Conversely, the independent development of audiences' interests and willingness to support novelty production due to larger demographic, socio-economic or cultural change may modify the authority of intermediaries providing them access to novelties and attempting to establish specific assessments, as was the case with the emergence of young audiences for the *Nouvelle Vague* in French cinema (see Chapter 5).

In the natural and human sciences, contributions to knowledge about shared objects of inquiry are primarily addressed to other members of their communities. The expansion of knowledge about these objects and the growing numbers of practitioners over recent centuries have driven the specialization of novelty production, with the shared objects of inquiry, the knowledge accumulated about them and the languages in which this knowledge was formulated becoming increasingly disassociated from the knowledge of even the scientifically literate who have acquired specialist degrees.

Scientific novelty has become relevant primarily to other specialist producers, who also are the only ones able to judge the relevance and technical quality of contributions so that the production of novelty depends on both a lengthy specialist education at universities and continuous involvement with the productions of other specialists. In Integrated communities where practitioners are directly dependent on each other's outputs for their own production of novel contributions, their authority is collectively exercised through the application of strong standards, while the exercise of epistemic authority is more differentiated in Polycentric communities where elective positioning and pluralistic productive relationships are more prevalent.

Given this concentration of authority in practitioner communities, they appear to completely control the allocation of opportunities to build protected space. However, an important background condition for this exercise of authority is that practitioner communities are maintained by intermediaries in science worlds. All communities of creative practitioners depend on intermediaries providing them with means of subsistence and with the means to produce novelties. Therefore, a minimum of authority always rests with non-practitioner audiences and intermediaries in science worlds, which must approve of the kind of novelty production in a very general sense in order to maintain them. Beyond this minimum of authority sharing intermediaries may exercise additional authority if there is a specific interest of society and thus intermediaries in the application of scientific findings.

Types of art worlds and science worlds

The ways in which intermediaries and audiences in art worlds and science worlds affect the selection of novelties for assessment, the assessment of

Table 4.1: Distribution of authority over aesthetic/epistemic purposes and standards and allocation of resources in modern arts and sciences

		Types of practitioner community	
		Polycentric	*Integrated*
Distribution of authority in art and science worlds	*Practitioner-dominated* (limited authority shared with intermediaries and audiences)	Many social sciences and humanities, some natural sciences	Most natural sciences
	Practitioner-led (some authority shared with intermediaries and audiences)	Technological social sciences and humanities (e.g. education, social work) Performing and multiply reproduced arts Painting and plastic arts producing single works	Technological biological and physical sciences (applied biomedical fields, engineering sciences, materials sciences)
	Intermediary-led (limited authority shared with practitioners and audiences)	Mass, heterocultural arts Investment-led arts	Industry-dominated engineering sciences (e.g. automotive engineering)

novelties and the support of particular productions of novelties depends on the distribution of authority in these worlds. Three major patterns of authority sharing between practitioner communities, audiences and intermediaries in art worlds and science worlds can be distinguished and be combined with the distinction between Polycentric and Integrated practitioner communities (Table 4.1).

In **Practitioner-dominated science worlds**, creative practitioners in some of the natural sciences, social sciences and humanities constitute the only audience that is competent to recognize novelties as relevant contributions and thus the only audience that exercises authority over this recognition. While the societal public is a general audience for creative contributions from these fields, it cannot participate in the appreciation of most of these contributions due to their esoteric character and context.

Since practitioners constitute the only audience relevant for the assessment of novelties in these worlds, major intermediary roles are also either executed by creative practitioners or dominated by them. The distributive role is executed by researchers taking on the roles of editors and reviewers of book series, books and journals. While publishers as non-practitioner intermediaries make essential contributions to their distribution, their decisions largely depend on assessments by members of scientific communities. Similarly, insofar the distribution of funds depends on assessments of quality and relevance, these assessments are made by members of scientific communities. With most major roles being executed by community members, the authority over standards, the recognition of contribution and decisions about further novelty production is largely concentrated in the practitioner communities, with only limited authority sharing with intermediaries and non-practitioner audiences. The distribution of authority within the practitioner community in these science worlds is polycentric in the social sciences and humanities as well as some natural sciences, or integrated in most natural sciences.

In **Practitioner-led art and science worlds**, creative practitioners share some aesthetic and epistemic authority with interpretative, distributive and resource providing intermediaries and variously involved and knowledgeable audiences. While research practices and aesthetic/epistemic standards in the more integrated applied sciences remain largely determined by practitioner communities, priorities and relevance criteria are here significantly influenced by audiences that are interested in using the novelties. Major audiences for such usable novelties are commercial interests and state actors, which can use the provision and allocation of resources to exercise authority over novelty production. These audiences and their intermediaries are also able to assess the utility of novelties for their purposes, which increases their authority over standards and practices.

Where creative communities in practitioner-led science worlds are polycentric, such intermediary and audience influence can extend to the support of particular epistemic approaches and standards. The development of quantitative empirical research projects on educational practices and structures in postwar Germany, for instance, was substantially supported by changes in regional governments' policies in the 1990s and 2000s. The German government and the federal states became interested in Germany's participation in international large-scale student assessments and addressed this expectation to the German educational research community, provided funding for this participation and provided researchers access to classrooms for empirical research (Gläser et al, 2014). The current state support for social scientific research on the conditions and consequences of climate change, digitalization and applications of artificial intelligence demonstrate that intermediaries may induce shifts in research priorities because practitioner communities share their authority over standards of relevance with intermediaries and society as the wider audience for scientific novelty production.

Many art worlds have Polycentric practitioner communities and are practitioner-led, with intermediaries and audiences exercising some authority over standards and practices. Intermediaries who interpret art works and control their diffusion to interested audiences, such as museum and exhibition curators, can exercise considerable authority over the recognition and adoption of art works, particularly when they share a common understanding of a field's purposes and development. However, just as creative practitioner communities vary in the extent of pluralism and heterogeneity of aesthetic approaches and standards, so too critics, art galleries, museums, publishers and other intermediaries differ in their adherence to common aesthetic standards.

In the modern arts, the recognition of a challenging new work as a relevant and technically competent novelty that could be worth developing into an innovation is often dependent on its appreciation by aesthetically informed intermediaries and audiences as well as by the specialist practitioner community. These influential groups sometimes include established artists in other fields, as when Emile Zola, Stephane Mallarmé, Felix Fénéon and other well-known writers commented positively on the work of Impressionist painters. As such works became more widely accepted, though, creative practitioners themselves were often able to gain support for further discontinuous innovations more directly. In the case of post-impressionism, for instance, Paul Signac and others published more practically focused accounts of their principles and purposes which probably had more impact on colleagues' novelty production (Bock, 1981, pp 13–21; Galenson and Jensen, 2002; Cottington, 2022, pp 136–42).

Furthermore, the distribution of aesthetic authority between different intermediaries can change. In his account of the demise of the avant-garde in post-Second World War American Art for instance, James Ackerman suggested that if the traditional role of:

> serious and imaginative (art) criticism is to bring to the notice of its public the distinguished but rejected art of its time … much of the critic's function has been pre-empted by curators and directors of museums and galleries, who exert more influence by exhibiting and publicizing new art than the critic can with words published in specialized journals. (Ackerman, 1969, p 381)

Since this postwar highlighting of the role of curators and public exhibitions in promoting new artistic approaches and styles, the 'figure of the curator has moved from being a caretaker of collections – a behind-the-scenes organizer and arbiter of taste – to an independently motivated practitioner with a more centralized position within the contemporary art world' (O'Neill, 2012, pp 1–2). Particularly with the expansion of large international group exhibitions, such as *Biennale*, since the 1960s, leading curators have become seen as performing more authorial/artistic functions in the development of contemporary art.

> Key here to the new museum curator's role in the planning of temporary 'ahistorical' exhibitions ... often arranged thematically or contextually ... in which the curator plays a role not unlike 'author' of the exhibition. The exhibition is a way to validate the originality of the curator's point of view, his or her aptitude for discovering new talents, and the artworks themselves by exhibiting them in a dialogue with each other to an initiated public. (Acord, 2010, p 449)

Just as art critics vary in the nature of the audiences they address and the ways they judge new art works, so too can galleries and other exhibition organizations perform different roles in presenting and supporting new artists and styles. In her account of the different ways that New York galleries promoted Abstract Expressionism in the late 1940s and early 1950s, Marcia Bystryn suggested they could be summarized in terms of two ideal types. The first:

> is characterized by its fostering of invention, allocation of symbolic rewards to the artist, personal ties with the artist, cultural goals, personnel who are artists themselves, and close connections with the artistic community. Type two is more involved with innovation than invention, rational economic goals, allocation of monetary rewards to the artist, business personnel, and close ties with the institutionalized art market. (Bystryn, 1978, p 394)

In the case of the Betty Parsons gallery, exemplifying the first type, the owner saw her primary role as a fellow artist fostering creativity rather than pursuing purely commercial goals and so being part of the creative practitioner community.

The influence of audiences on aesthetic goals and standards in practitioner-led art worlds has usually been exercised indirectly through attendance at exhibitions and performances or acquisitions of books and other multiply reproduced outputs. In the case of unique paintings and sculptures, though, some audience groups are able to exercise authority over aesthetic reputations and values in particular circumstances. As Impressionist practices and purposes became more widely acceptable and collected in France towards the end of the 19th century, for example, collectors' interest in the works of Cézanne, Gauguin, van Gogh and other Post-Impressionists grew and encouraged the *denicheurs*, or winklers-out, of contemporary art to invest in unconventional works by largely unknown artists in the early 20th century (FitzGerald, 1995, pp 20–6).

The financial success of such speculative investments was dramatically demonstrated by the public auction of the collection of mostly 20th-century paintings acquired by Andre Level on behalf of the *Société de la Peau de l'Ours* (Bearskin Group) in March 1914 (Cottington, 2022, pp 120–2). This group of 14 collectors was established in 1904 to acquire 'important works by young painters or those just establishing some notoriety' (Cottington, 2004, p 14) from makeshift galleries, enthusiasts' flats and artists' studios for sale ten years later. As Michael FitzGerald emphasizes:

> The scale of their purchases rivalled that of better-known patrons like the Americans Leo and Gertrude Stein and the Russian Sergei Shchukin, but in the diversity of its interests La Peau de l'Ours surpassed these collectors ... Moreover, the group's frequent practice of buying directly from artists reveals not only the broad pattern of collecting but also the effect of individual exchanges on these artists' work and careers. (FitzGerald, 1995, p 16)

Such direct involvement in the acquisition of new works on a substantial scale encouraged other collectors as well as critics to become interested in the Parisian avant-garde and dealers, such as Vollard and Kahnweiler, to invest in their artistic careers.

Considering finally the distribution of authority in **Intermediary-led art worlds and science worlds**, practitioners here exercise only limited collective authority over the practices involved in producing novelties and the standards for evaluating novelties. In the arts, this category includes most fields where new works are produced to appeal to the largest possible audiences and are mass reproduced for this purpose (Crane, 1976; Carroll, 1998, pp 184–211). In these fields, intermediaries involved in the production and distribution of art works can become the dominant authoritative agency determining the acceptability of novelties and the allocation of resources. In particular, the commercial enterprises using their income from selling art works to such audiences to fund the production of new works often determine the conditions for the production of novelties, as in the heyday of the Hollywood studio system (Gomery, 1996; Schatz, 1996; Baumann, 2007, pp 76–82) and during the oligopolistic periods of the US record industry (Peterson and Berger, 1975).

The concentration of authority with intermediaries crucially depends on their control of practitioner's access to audiences funding the production of novelties. As two examples of innovations in intermediary-led art worlds demonstrate, changes in audiences and the disruption of intermediaries' control may weaken the authority of incumbent intermediaries and make room for new intermediaries gaining authority. In the case of Bebop in modern jazz, the initial access for Black musicians to small audiences in clubs in Harlem and the later temporary weakening of oligopolistic record companies' control over the recording and marketing of new music gave authority to intermediaries controlling access to new audiences (those interested in Jazz as concert music). Similarly, in the case of the *Nouvelle Vague* in French cinema the French state provided funding for unusual and experimental documentaries and feature films and thus temporarily acted as an intermediary that enabled new practitioners to develop novel aesthetic approaches and interests that appealed to new young audiences (see Chapter 5).

A second type of intermediary-led art worlds are the investment-led arts. In the contemporary art market, intermediaries such as gallerists, critics, museums and international exhibition organizers have come to share aesthetic authority and resources with the relatively small audience of rich collectors, auction houses, managers of art fairs and investors in collective art funds who are often indifferent to their aesthetic qualities. While some internationally renowned artists have been able to influence this market, it is unclear that the community of creative practitioners is collectively able to affect, let alone control, aesthetic standards and relevance criteria (Gnyp, 2020; Buchholz, 2022). We shall briefly discuss the implications of this development in the modern visual arts in Chapter 6.

Intermediary-led science worlds differ from other science worlds in that an external audience of commercial clients replaces the community as primary audiences. They exercise considerable authority over the community's goals and standards, sometimes in collaboration with intermediaries for the allocation of funding which they control. Communication between creative practitioners is constrained by secrecy and intellectual property rights, and relationships to paying clients dominate over productive relationships in the community. This authority distribution has been observed in automotive engineering in Canada and Germany (Rutherford and Holmes, 2008; Grieser, 2018).

These six types of art worlds and science worlds differ in the ways in which protected space for the development of innovations is granted or can be built. In practitioner-dominated science worlds, these opportunities mainly depend on unspecific public support for the community's work on its epistemic problems. If and insofar as this support exists, the differences between Integrated and Polycentric practitioner communities as environments for the development of different kinds of innovations discussed in the previous section are not affected by the authority of audiences beyond the practitioner community itself or of intermediaries.

In practitioner-led art and science worlds, building or granting protected space further depends on intermediaries supporting it by valuing particular novelties and mobilizing resources for the development of innovations. In the case of Polycentric practitioner communities, additional selectivity of intermediary support may exist if intermediaries selectively promote particular approaches to novelty production.

In intermediary-led art worlds and science worlds, building or being granted protected space depends on the mobilization of audiences for the provision of resources. These processes are largely independent of practitioner communities being integrated or polycentric because securing access to audiences through particular intermediaries is more important for practitioners than the support of their communities.

5

Developing Different Kinds of Innovations in Different Art and Science Worlds

Developing innovations in different kinds of art and science worlds often involves changes in established authority and productive relationships. This is particularly so when they imply substantial shifts in productive practices and aesthetic/epistemic standards and purposes. In this chapter we summarize and compare the development of nine process, complementary and displacement innovations in differently organized art and science worlds in order to identify the conditions that enabled and supported their changes in particular societal contexts. The number of case studies discussed here is limited by the availability of empirically based accounts of the changing practices and purposes of creative practitioner communities – as distinct from those that focus primarily on the works of individuals – and their broader social environments (Table 5.1).

Considering first the development of innovations in intermediary-led fields, we compare the nature and context of two significant innovations in the arts. Initially, the creation and widespread adoption of the Bebop style in modern jazz in the 1940s as a displacement innovation, and then the production of the *Nouvelle Vague* style of film making in French cinema between 1959 and 1962 as a complementary innovation.

Next, we consider four cases of innovation development in practitioner-led fields of the arts and sciences. Initially, the creation and rapid diffusion of Cubism as a displacement innovation in painting and sculpture, then the production and adoption of metafictional styles in novel writing as a complementary innovation in Anglophone fiction; third, the change in the dominant approach to studying cancer from searching for a viral agent to focusing on genetic mutations as a displacement innovation in the sciences; and fourth, the transformation of business finance research and teaching into Financial Economics as a displacement innovation in the social sciences.

Table 5.1: Distribution of case studies in different art and science worlds

		Types of practitioner community	
		Polycentric	*Integrated*
Distribution of authority in art and science worlds	*Practitioner-dominated*		Development of Radio Astronomy as a process innovation. Experimental realization of Bose-Einstein Condensation as a process innovation. Evo-devo as a complementary innovation.
	Practitioner-led	Cubism as a displacement innovation. Metafiction as a complementary innovation. Oncogene Approach as a displacement innovation. Financial Economics as a displacement innovation.	
	Intermediary-led	Development of Bebop style as a displacement innovation in jazz. Development of *Nouvelle Vague* style of film making as a complementary innovation in France.	

Finally, we compare three cases of innovation development in the practitioner-dominated sciences. First, the development of Radio Astronomy as a process innovation in the 1940s and 1950s; second, the successful experimental realization of Bose-Einstein Condensation (BEC), which began as a process innovation in the 1990s and later also became a complementary innovation in the 2010s; and third, the development of Evolutionary-Developmental Biology (Evo-devo) as a complementary innovation in the biological sciences.

The presentation of each case study includes descriptions of the nature of the innovation, its development in practitioner communities and the conditions facilitating this development as far as they could be extracted from the primary literature. We also applied the analytical categories developed in the previous chapters and identified the kind of innovation and the structure of the practitioner community developing it. These categorizations were problematic in some cases because the character of some innovations changed when they were developed and because the development of the innovations changes the structure of practitioner communities. Where such changes could be derived with sufficient certainty from the literature, we included them in the case studies.

We then briefly consider two instances of novelty production that initially attracted some creative practitioners' attention but later failed to generate sufficient collective interest and commitment to develop further into a significant community innovation, namely cold fusion and aleatory fiction.

Innovation development in intermediary-led art worlds

a) Bebop as a displacement innovation in jazz

The nature of the innovation

The development and success of the Bebop movement in modern jazz have been considered to be analogous to paradigm changes in the natural sciences. For example, in his reconstruction of the birth of Bebop, Scott DeVeaux has suggested that:

> Through the transformative example of Charlie Parker and Dizzy Gillespie, jazz in the 1940s experienced a 'reconstitution of the field from new fundamentals'. All the familiar symptoms of revolution, the world turned upside down, are there: the youth of the revolutionaries, the startling, unexpected nature of the new insight; the specter of an older generation clinging to the old paradigm, even as hordes of new practitioners rush to embrace the new; and finally, the triumph of the new paradigm and the recasting of the new field in its image. (DeVeaux, 1997, p 43)

Similarly, in contrasting Bebop with the mainstream Swing jazz, Thomas Owens has emphasized the revolutionary nature of Bebop drumming and the role of the rhythm sections in performing modern jazz:

> bebop rhythm sections, using on- and off-beat chordal punctuations (known as comping) supplied by pianists and guitarists, and additional punctuations supplied by the drummer on drums and cymbals, were much more polyrhythmic than were swing rhythm sections. Also, bebop harmonies were more dissonant, and the melodies improvised upon these harmonies were more chromatic and less symmetric rhythmically. Bebop themes … often were more chromatic, disjunct, and rhythmically complicated. (Owens, 1995, p 4)

This novel musical language:

> with its strange-sounding notes and breakneck tempos … was everything swing was not. Its explosive excitement centred on individual musicians, who took turns playing improvised solos. Often Bebop was performed at a pace far too fast for dancing. [It] was intended to be played by small groups that joined together only briefly at the beginning and end. Bebop was so intricate that most musicians – even experienced jazz musicians – could not imitate it with ease … The secret of the new music rested in soloing musicians' interpretations of the song's chords – they added notes in some cases and left out notes in others. Speed, improvisational risk taking, and on-the-fly harmonies also played big roles in the music's high-wire act. (Myers, 2013, pp 11–2)

In sum, Bebop was performed by small groups of musicians at a fast pace with extensive improvisations, considerable harmonic complexity and frequent rhythmic changes.

Developing the innovation

Bebop was 'first established in the early and middle 1940s in New York City by Charlie Parker, Dizzy Gillespie, Thelonius Monk, Bud Powell, Kenny Clarke, Max Roach, Ray Brown, and a few others' (Owens, 1995, p xviii) who had begun to perform professionally between 1935 and 1942. Often working in Black touring bands performing swing music primarily aimed at dancing audiences, usually Black, they became part of the national commercial entertainment industry, albeit in subordinate and unstable roles. As the popularity of this kind of music began to decline and competition for work intensified in 1939 and 1940, particularly for Black bands, many sought more stable jobs in New York City which had become the national cultural capital and centre for jazz performances (DeVeaux, 1997, pp 153–7). Although many audiences there remained segregated, the variety of employment opportunities enabled Black musicians to combine performing established musical styles to all-white clubs and theatres with exploring new styles with fellow musicians in the Harlem after-hours clubs (Stump, 1998). Two of these clubs have become particularly associated with the development of Bebop music: Minton's Playhouse and Clarke Monroe's Uptown House (DeVeaux, 1997, pp 219–34).

These clubs were attractive to ambitious musicians because they delegated considerable autonomy to the house band and proficient visitors in deciding what to play and whom to play with in competitive jam sessions. By fostering a sense of exclusivity and technical excellence, they:

> helped to shape bebop's cultural identity as an arcane form of hip musical expression, beyond the understanding and capabilities of musicians outside the community of innovators who created it. Its association with Harlem also helped to establish bebop's identity as an African-American innovation. Indeed, one of the often cited motives of bebop's black creators was to develop a style of jazz which, unlike swing, could not easily be appropriated by more commercially successful white musicians. (Stump, 1998, p 27)

This style was generally more complex than the dance oriented swing style, with more dissonant harmonies and improvised melodies that were more chromatic and rhythmically complicated. Bebop arrangements often combined established chord progressions and themes with novel melodies in melodic contrafacts, typically consisting of 'an improvised piano introduction, a theme chorus (played in unison by groups containing two or three wind instruments), a series of improvised solo choruses, and a closing theme statement' (Owens, 1995, pp 4–5). They were usually played by much smaller groups of musicians than the large swing bands (quartets, quintets, sextets) and at considerably faster speeds.

A significant change in the role of the rhythm section, however, was initiated by Kenny Clarke when he was playing in Teddy Hill's big band in the late 1930s. He stopped:

> automatically playing the bass drum on every beat and began playing a variety of on- and off-beat punctuations on the bass drum and snare drum ('bomb'). Moving his right hand from the high hat (situated on his left) to his ride cymbal (on his right) gave him more room to maneuver his left hand on the snare drum (directly in front). (Owens, 1995, pp 9–10)

While this novel technique of 'keeping a shimmering pulse on the ride cymbal, occasionally punctuated by vigorous accents on the bass and snare drums … was greeted enthusiastically by some younger musicians … it angered one of Hill's veteran trombonists, who complained to Hill' (DeVeaux, 1997, p 218) and had Clarke fired in 1940. Shortly thereafter, Hill became the manager of Minton's Playhouse and appointed Clarke as the house band leader, encouraging him to continue dropping all the bombs he wanted to (DeVeaux, 1997, p 219).

The exclusivity and relative isolation of the Harlem clubs limited the commercial opportunities for Bebop musicians to gain the attention and support of larger audiences, particularly as Black musicians had been mostly excluded from performing on prime-time radio shows and in prestigious hotel ballrooms across the country (DeVeaux, 1997, pp 146–51). However, by the end of the 1930s the two blocks of 52nd Street West of Fifth Avenue had become a major centre of jazz performance in New York City, as part of a larger entertainment district across midtown Manhattan. While this concentration of nightclubs featuring live jazz had developed during the swing era, it did provide opportunities for innovative jazz styles to attract relatively large audiences of jazz enthusiasts once they had demonstrated some success. In the case of Bebop, this occurred after it had developed into a more recognizable (and marketable) style in the mid-1940s (Stump, 1998, pp 27–9). As DeVeaux has emphasized:

> The commodification of bebop thus required a conscious commitment … to reorient the music to existing commercial channels. This process was set in motion early in the 1940s by the lure of public jam sessions and the night clubs on 52nd Street, both of which put jam-session-style jazz on a profitable footing … Along the way the music would have to be deliberately transformed. The repertory would have to be converted into clearly defined economic units, preferably original compositions, for which authorship could be precisely established. (DeVeaux, 1997, p 298)

By the late 1940s Bebop had become:

> the dominant jazz style, thanks to the perseverance of musicians, jazz writers, concert promoters, disc jockeys, and agents – all of whom gained visibility and profited from bebop's rise. Some five years after the first recording of bebop, in February 1944, the new music was being played on the radio in nearly every urban market. (Myers, 2013, p 46)

However, as larger numbers of musicians deciphered its structure and Bebop became the new sound of mainstream jazz, it lost its revolutionary qualities and increasingly served as the common foundation of new jazz styles (Gioia, 2011, pp 253–85; Myers, 2013, p 47).

Conditions supporting the development of the innovation

The need to establish ownership of property rights by being recognized as the composer of new music rather than simply its performer involved a change in the traditional status and role of jazz musicians as performers of standard songs for dancing audiences. This was facilitated by the creation of the Broadcast Music Incorporated (BMI) performance rights organization by the radio networks at the beginning of the 1940s as a way of breaking the American Society of Composers, Authors and Publishers (ASCAP) monopoly control over these rights. In building its catalogue quickly, the BMI 'turned to grass-roots sources previously excluded from ASCAP – folk music, hillbilly (as country music was then known), Latin American music, jazz – and it remained committed to these marginalized genres after it was securely established' (DeVeaux, 1997, p 302). This enabled innovative jazz musicians, such as Dizzy Gillespie and Thelonius Monk, to become established as arrangers and composers of new music as well as its elite performers and encouraged the production of original compositions.

By the late 1930s the three major record companies, Decca, Columbia and Victor, that collectively produced and distributed 85–90 per cent of the market, had become more aware of the growing demand for small-combo jazz, but the combination of increasing defence spending, record sales of hit recordings in 1941 and the shortage of shellac for producing 78 rpm records following the outbreak of war at the end of that year considerably reduced their willingness to support small bands playing novel music (DeVeaux, 1997, pp 298–9; Myers, 2013, p 14). As a result, it became very difficult for musicians without current record contracts to reach audiences beyond live performances in small nightclubs and similar venues. Subsequently, recording became virtually impossible in August 1942 when the American Federation of Musicians (AFM) imposed a national ban on members making records. This was intended to force the record companies to pay a royalty on the sale of records to a union trust fund that would subsidize live music by hiring musicians left unemployed by the rising use of records by radio stations and jukeboxes (Myers, 2013, pp 5–6). It resulted in the cessation of new music recording and distribution.

However, within a few months of Decca agreeing to the AFM's terms in September 1943, largely because it lacked the backlog of recordings that Columbia and Victor owned, 76 newly created record firms were established and agreed to these conditions. Since many of the popular swing jazz bands were already signed to the big three companies, these new firms turned to smaller jazz combos for material, including some of the Bebop pioneers (Myers, 2013, pp 23–6). The audience for jazz records

> was small but dedicated with a disproportionate representation of high school and college-age consumers, who were willing to pay premium prices for records that satisfied their tastes. Recording artists could be easily drawn from a pool of skilled freelancers working on 52nd Street, eager for a chance to record and willing to work for union scale. The flexible jam-session-style performance format precluded the need for expensive arrangements. (DeVeaux, 1997, p 303)

Furthermore, when the ban on making records was removed across the industry, as RCA Victor and Columbia accepted the AFM's terms on

Armistice Day in 1944, the union removed the ban on radio stations using records on the air. This helped to expand the role of disc jockeys on radio stations who:

> chose which records to play, introduced the music to the audience, and educated listeners, explaining to them in colorful language why a specific record or artist was exceptional … If a disc jockey was successful in creating an appetite for an artist's music, audiences would play the records on jukeboxes and buy the discs at record stores. Most important, the disc jockey was the ever-important hub for the record companies, concert promoters, jazz critics, and the listening public. (Myers, 2013, pp 36–7)

By the end of 1948, Bebop had become the dominant jazz style and a distinctive jazz art world supporting the idea of modern jazz as a music genre and aesthetic separate from popular music was firmly established. As Lopes (2002, p 216) has suggested:

> This art world of concerts, clubs and record labels also formed the basis of the spread of the modern jazz paradigm among professional musicians as well as developing an audience for modern jazz. As a decentralized commercial market, the jazz art world provided an open system of production that allowed different genres of jazz to find a niche within this market.

b) The Nouvelle Vague *as a complementary innovation in French cinema*

The nature of the innovation

The technical and aesthetic changes in film making practices and purposes developed by the 'New Wave' of over 100 directors making their first feature film between 1959 and 1962 in France constitute an example of a complementary innovation in an intermediary-led art world. In contrast to the prevalent studio-based 'tradition of quality' films that focused on faithful adaptations of successful plays and novels and costume dramas of historical events with well-known actors and followed a strict, pre-established shooting script, these film makers worked with young actors in making films that dealt with the lives and problems of younger audiences with unusual techniques and approaches based on the distinctive qualities of the cinema (Rachlin, 1993). Technical novelties included the frequent use of location shooting with handheld cameras and sound recorders in natural light, frequent improvisation during filming, elliptical editing, obtrusive camera movements, variegated film speeds, sudden shifts in shot distances and jump cuts in order to disrupt the temporal and spatial continuity of the traditional narrative (Marie, 2003, pp 70–87).

Intended to emphasize the self-reflexivity of film making and the dominant role of the director as the 'author' of the film, such technical disruption was 'complemented by dramaturgical experiment, with loose causal connections, disconcerting shifts in tone, digressions, ill-defined character motivations and ambiguous conclusions, all reinforcing the aesthetic distance between the viewer and the film' (Parkinson, 2012, p 186). As Mark Cousins comments about the numerous jump cuts in Jean-Luc Godard's *A Bout de Souffle* (1959):

> The reason for cutting in this way was because the cuts were beautiful in themselves, because they emphasized that what we were watching was cinema, just as painters had turned to Cubism many years earlier because it emphasized the flatness of the canvas ... movies were not just vehicles to carry stories and information. (Cousins, 2011, pp 270–1)

These preferences were intended to provide greater flexibility for film directors by weakening the heavy constraints typical of the commercial and industrialized cinema model and 'erasing the borders between professional and amateur cinema, and those between fiction, and documentary, or investigative films' (Marie, 2003, p 71).

Developing the innovation

The *Nouvelle Vague* was largely developed by two groups of directors making their first feature film between 1959 and 1962 who would constitute the elite of French cinema for the next 20 years (Rachlin, 1993). The first is commonly associated with the critics attacking the mainstream tradition of quality films in specialist magazines such as *Cahiers du cinéma* in the 1950s, particularly Francois Truffaut *(Les Quatre Cents Coups* 1959), Claude Chabrol (*Le Beau Serge* 1959 and *Les Cousins* 1959), Eric Rohmer (*Le Signe du Leo* 1960), Jacques Rivette (*Paris Nous Appartient* 1960 and Jean-Luc Godard (*A Bout de Souffle* 1960). These young directors were:

> all *cinéphiles*, or very devoted film buffs, who knew their film history inside out. They acquired their understanding of cinematic culture from watching thousands of films, gaining clear ideas about *mise-en-scène* that were based on aesthetic choices, moral options, personal tastes, and sometimes violent disapproval. (Marie, 2003, p 26)

Many of these continued to make critically and commercially successful films for the following decades, albeit some, such as Chabrol, became less stylistically innovative. Godard in particular 'has devoted each of the five distinct phases of his career redefining that relationship (between fiction and reality) in order to exploit cinema's potential for intellectual, political and artistic expression' (Parkinson, 2012, p 188).

Another group was more linked to political and 'Left Bank' themes and the left-wing film magazine *Positif*, as well as to some of the complex and self-reflective novels of the 1950s (Graham, 1996, p 578). These included Alain Resnais (*Hiroshima, Mon Amour*, 1959, and *L'Année Dernière a Marienbad*, 1961), Agnes Varda (*La Pointe Courte*, 1954) and Chris Marker, who focused mostly on making documentaries, such as *Cuba, si!* (1961), *Le Jolie Mai* (1963) and *Far from Vietnam* (1967). Many of these films dealt with the manipulation of time and the paradox of memory through stylized narratives, notably *L'Année Dernière a Marienbad* in which this preoccupation 'inspired such characteristic technical traits as the elliptical transition between the objective and subjective narrative modes and (Resnais') use of stylized tracking shots to present the past, present and future upon the same spatial and temporal plane' (Parkinson, 2012, p 191).

Notwithstanding their thematic and stylistic diversity:

> the film-makers of the French New Wave shared a desire to demolish the conventions that had sustained the narrative film since Griffith and replace them with an audiovisual language sufficiently eloquent and malleable to provide anyone working within the art form with a rich and powerful means of self-expression. Such was the success of their undertaking that film industries worldwide felt the impact of the nouvelle vague, many experiencing new waves of their own. (Parkinson, 2012, p 193)

A common focus of many New Wave films concerned the problems and interests of younger audiences in France, often using young non-professional actors to explore the themes that had been celebrated in the commercially successful novels of Francoise Sagan. As Rachlin has suggested:

> Sagan's role in setting the stage for the *Nouvelle Vague* is threefold. First, she proved the viability of the young as consumers of a differentiated cultural product. Second, she demonstrated that the young individuals, themselves, might be best suited to produce this product. And, finally, third, she showed the clear commercial appeal of the kind of themes – sexual liberty, moral ambiguity and self definition through consumption – that men like Truffaut and Godard would later bring to the screen. (Rachlin, 1993, p 438)

Conditions supporting the development of the innovation

There were two major sets of conditions supporting the rapid development and impact of the *Nouvelle Vague* in French cinema at the end of the 1950s.

First, the promotion and variably successful establishment of an art-house audience for artistically ambitious films from the 1920s onwards, together with associated critical journals dedicated to the promotion of film as an artform. Second, the substantial level of state support for the film industry in the 1950s and 1960s, including the production, distribution and exhibition of unusual and challenging films (Marie, 2003).

Despite the limited size of the French film-going population in the 1920s and 1930s compared to those in Britain and Germany, the formation of an informed intellectual audience for films was more active in France than elsewhere. As Colin Crisp has pointed out, although:

> the classic French cinema never succeeded in fashioning the mass audience it aspired to, it was extremely effective in developing mechanisms for fashioning an art-house audience. This process began much earlier and was much more systematic in France than elsewhere. Basically the audience concerned was the middle-class intellectual and cultural elite; they saw themselves as working for the defense of French cinema and of cinema as a cultural form, and their means of operation were the ciné-club, congresses and study groups, the concept of the cinématique, a separate theater circuit committed to art films, and critical journals of a more reflective kind. (Crisp, 1993, pp 226–7)

Ciné-clubs and critical journals devoted to the cinema were the longest-lived and more effective of these mechanisms for developing an art-house audience for aesthetically interesting films from the mid-1920s onwards, albeit with some turnover of individual organizations. In the 1930s there were around 20 *ciné-clubs* in France, mostly in the Paris area, that presented avant-garde films in a tripartite format of an initial verbal account, the projection of that film, and a general debate and discussion.

During the Occupation, *ciné-clubs* as such were completely forbidden but grew dramatically from one in November 1944 to about 200 in the 1950s and 235 in 1960, distributed in cities and towns across the country. While the postwar audiences of 60,000 to 80,000 annually in the 1950s, reaching 100,000 in 1960, mostly from liberal and bourgeois occupational categories, were too limited to make much impact on the commercial cinema, they provided a forum for *ciné-files*, critics and would-be film makers to see and discuss French films and foreign-made ones that were often subjected to censorship restrictions. Additionally, the postwar *Cinémathèque française* enabled access to the history of world cinema and a location for ambitious practitioners such as the *Nouvelle Vague* directors to meet and plan their contributions (Crisp, 1993, pp 227–32).

The second major factor in the rapid development and success of the *Nouvelle Vague* as a collective aesthetic movement at the end of the 1950s was the substantial postwar state support of the national film industry, including funding for upgrading and extending the smaller cinema chains (Crisp, 1993, pp 74–9; Marie, 2003, pp 53–61). The state also continued many of the organizational reforms introduced during the German occupation that rationalized the film industry. These included the banning of double feature film programmes to save film stock and the provision of support for making short films and documentaries that provided a training ground for young filmmakers to explore and demonstrate their competences. Changes to the system of state support for the film industry in the 1950s introduced the criteria of the quality and likelihood of providing new perspectives on cinematic art in awarding financial support for new films for the first time, which helped the production of the first films of Chabrol and Truffaut (Marie, 2003, pp 54–5).

Overall, then, as Crisp insists:

> if a number of producers continued to invest in high-risk artistic ventures during the difficult post-war years and on into the sixties, it was not least because their risks were largely underwritten by a state subsidy system aimed specifically at fostering ambitious artistic experiments … these subsidies were ultimately ploughing half a billion francs each year into projects that foreshadowed the New Wave and indeed that nurtured it. (Crisp, 1993, p 417)

Additionally, the combination of: (a) rapidly declining attendances at cinemas, particularly among the old and middle-aged, as televisions became more widely available, (b) increasing pressures to reduce production costs and (c) the emergence of distinctive consumption patterns among younger groups encouraged the adoption of flexible and cheaper film making practices. These included location shooting with faster film stock and lighter cameras and sound equipment, to make films that appealed more directly to young people (Rachlin, 1993; Graham, 1996, pp 576–7; Cousins, 2011, p 268).

This willingness of investors to support films aimed at the expanding youth-based subculture was considerably encouraged by the commercial success of the American *East of Eden* and *Rebel Without a Cause* in 1955 and Roger Vadim's *Et Dieu Créa La Femme* in 1956 (Rachlin, 1993). Despite the greater costs of producing and distributing films than those involved in the development of novelties in other artistic genres, such as paintings and literature, innovative filmmakers and their producers in France were able to obtain both private and public support for making technically unusual and aesthetically novel films.

Summary

These two case studies of innovation development in intermediary-led art worlds illustrate the importance of changing societal contexts in enabling and supporting them. In both cases the authority of intermediaries over the production, distribution and presentation of new products was weakened relative to that exercised by innovative groups of practitioners developing new styles and reaching new audiences. In the case of Bebop, this resulted from the combination of the AFM ban on recording new music in 1942 and the subsequent establishment of new companies willing to record new jazz styles, the creation of the BMI performance rights organization, the difficulties of touring during the war – particularly for Black musicians – and the expansion of the audiences for listening to live performances in urban jazz clubs and concert halls. The continuing promotional efforts of disc jockeys, concert promoters, musically informed critics and musicians' agents together with the rapid expansion of national radio networks and jukeboxes in the mid to late 1940s helped to establish Bebop as the new sound of mainstream jazz.

In the case of the French New Wave, changes in the production, financing and exhibition of feature films combined with shifts in the nature of audiences to enable innovative artists to gain greater authority over film making and interpretation. Postwar state support for the French film industry created opportunities for young directors to gain experience in making short films, particularly documentaries, and its later focus on innovative projects encouraged producers to invest in new styles of film making involving greater improvisation and control by directors. As the mass market for more traditional approaches declined in the late 1950s, the success of both American and French films addressed to the concerns and lives of more youthful audiences provided additional encouragement for investing in innovative directors who could appeal to this audience. Further support for the idea of directors being the controlling authority over film making rather than script writers or producers had, of course, come from many *ciné-file* critics since the 1940s.

Innovation development in practitioner-led art and science worlds

a) Cubism as a displacement innovation in the plastic arts

The nature of the innovation

Cubism was an artistic movement that initiated a pictorial revolution in the ways that images were constructed and presented. Generally understood to have been initiated by Georges Braque and Pablo Picasso between 1907 and 1914, it rejected the traditional techniques of: (a) *chiaroscuro*, or using light and shade to create the illusion of volume, (b) foreshortening objects to imply depth and (c) portraying subjects from a single viewpoint and perspective, as well as the assumption that paintings must present an illusion of reality at a particular moment in time. Instead, Cubists constructed a new pictorial language and aesthetic that involved the construction of a painting in terms of a linear framework, the fusion of objects with their surroundings, the combination of several views of an object in a single image, and of abstract and representational elements in the same picture in order to emphasize the flat, two-dimensional nature of the canvas and the many and varied ways of seeing and understanding the world (Fry, 1966, pp 9–11; Golding, 1968, p 27; Cox, 2000, pp 11–3; Antliff and Leighten, 2001, pp 7–10).

Central to this new aesthetic was a reconceptualization of three distinct tensions or problems in painting. As summarized by Michael Baxandall (1985, pp 44–6) these were: (1) how to represent the three-dimensional nature of objects and people on a two-dimensional canvas without creating a mere illusion of depth, (2) how to resolve the conflicting priorities of form and colour, and (3) how to deal with the tension between representing a moment of experience on the canvas and the extended process of thinking about and constructing images of experiences.

This concern with some of the basic problems inherent in all painting since the Renaissance highlights the intellectual focus of the Cubism of Braque and Picasso and its contrast with the more expressionist intentions of Fauvism and much of the avant-garde art of the previous 50 years. As John Golding (1968, p 60) has emphasized:

> the Cubism of Picasso and Braque was to be essentially conceptual … their paintings are not so much records of the sensory appearance of their subjects, as expressions in pictorial terms of their idea or knowledge of them … it was the conceptual element of Cubism that enabled the painters at various times to detach themselves from visual appearances without losing touch with the material world around them.

Developing the innovation

An early novelty that became seen as heralding one of the most important formal features of Cubist painting was the treatment of the figures in Picasso's *Les Demoiselles d'Avignon*, considered by Edward Fry as 'probably the first truly 20th century painting … (it) contained new approaches both to the treatment of space and to the expression of human emotion and states of mind' (Fry, 1966, p 12). In reducing these subjects to simple, angular shapes or planes arranged in shallow depth that suggested the forms of solid objects and their relationship to the space around them, Picasso combined different views of them into a single image that defied the system of single-point perspective and light source that had dominated Western European art since the Renaissance (Golding, 1968, pp 57–8). After encountering this painting in November 1907, Braque:

> needed little reflection to understand the significance … of Picasso's wilful distortions of figures and ambient space: within days he was working on a multi-figure composition of his own, which transcribed these figures … of the *Demoiselles* into the terms of his own concerns. The resulting painting, *Woman*, exhibited at the (*Salon des*) *Independents* the following March. (Cottington, 2004, p 35)

Another compositional device that further distanced Cubist paintings from illusionist practices was the use of a grid of lines, shading and textures to organize lines and planes into 'an uneven rhythm of light, shade and line tentatively aligned to the rectilinear edges of the canvas' (Cox, 2000, p 126). Such grids were more explicitly expressed in the work of Juan Gris who joined Picasso and Braque in their development of Cubism in 1911. He used vertical and horizontal lines to form a linear framework that divides figures and their settings into distinct sections in which the subject is studied from different points of view (Golding, 1968, p 100).

A very striking novelty that led to a whole series of developments in 20th-century art from 1912 onwards was the attachment of extraneous objects or materials, such as oil cloth, wallpaper and mirrors, to the picture surface. These *collages*, or glued substances, and *papier-collés* (pasted papers) emphasized the weight and solidity of Cubist pictures as constructed material objects quite different from conventional oil paintings and opposed to the idealized conceptions of works of art. They also helped to transform the art of sculpture in the 20th century (Golding, 1968, pp 103–5; Cox, 2000, pp 261–2; Cottington, 2004, pp 122–31).

While Braque and Picasso tended to use *papier-collés* as fragments of coloured shapes suggesting particular objects as well as being pieces of extraneous matter, Gris began to cover the entire surface of the canvas with cutouts of variously coloured paper to achieve:

> unusual compositional effects of great subtlety and of almost knife-like balance … what distinguishes Gris' original approach to *papier collés* from that of Braque and Picasso is that he used it primarily to give his work a sense of definition and certainty, a precision and exactitude that he could not achieve by any other means. (Golding, 1968, pp 110–1)

Picasso, and later Braque, also differentiated themselves from most painters in Paris by not seeking legitimacy and acclaim through the public display of their works in the Salons to reach audiences and collectors. Particularly after the production of *Les Demoiselles d'Avignon* in 1907, which was not publicly exhibited until 1916 at the *Salon d'Antin* (Cox, 2000, p 78), Picasso stayed aloof from these established means of gaining attention and support in France, relying instead on the commitment of private galleries and a small group of collectors (Cox, 2000, p 84; Cottington, 2004, pp 14–8). Despite this seclusion from the public French art world, their early Cubist works were seen and discussed by a limited number of other artists, such as Fernand Leger and Robert Delaunay, who developed their own responses that were exhibited in 'high profile shows in the coming years (and) helped to make it (Cubism) into a household word in pre-war Paris' (Cox, 2000, p 132).

Conditions supporting the development of the innovation

This production and successful diffusion of radically novel approaches to painting and sculpture were facilitated by three major changes in the French, and more broadly European, art world at the end of the 19th century. First, shifts in the market for new art works and styles that provided protected space for radical innovators; second, intensified competition between the expanding number of artists in Paris for public attention and support, together with the expanding variety of media for gaining audiences and legitimacy; and third, increasing diversity of aesthetic and economic authority over aesthetic standards and priorities in France and elsewhere.

The rise in prices for paintings by the impressionists in the 1890s, partly due to the entry of American collectors into the market, was followed by the rapid increase in prices of works by the post-impressionists in the 1900s, especially those by Cézanne, which encouraged the expansion of speculative investments in works by younger, unorthodox and relatively unknown artists. In 1904, for example, the art-collecting syndicate *Société de la Peau de l'Ours* (Bearskin group) was set up by Andre Level to invest an annual sum of money in buying works of young painters over a ten-year period to be sold at auction at the end of that period, in the hope (achieved in this case) of realizing a substantial profit (Cottington, 1998, pp 44–5). Such *denicheurs*, or winklers-out of bargains, focused on finding interesting novelties being shown in the smaller private galleries and artists' own studios rather than the kinds of works being displayed in the Salons. Collectively they bought enough such paintings from these dealers for them to promote artists producing experimental works, notably Daniel-Henry Kahnweiler who had rented a small gallery space near the Place de la Madeleine in 1907 and pursued a relatively novel business strategy in supporting Braque and Picasso:

> He bought directly from the artists, introducing the innovation of 'exclusivity', that is, an agreement to take every picture the artist

> produced, at set prices according to size. This was a device unheard of until then in the cautious world of the Paris market ... Kahnweiler accompanied this selectivity with an exhibition policy that combined inaccessibility in Paris – requiring his artists to refrain from showing anywhere else, salons included – with an acceptance of every possibility opportunity of foreign exposure. By means of such entrepreneurship he also accumulated a small but loyal clientele of *denicheurs*, ... for a select, but internationally recognised, stable of artists. (Cottington, 2004, p 15)

Such commitments provided the support for Braque and Picasso to concentrate on producing relatively small-scale pictures of still-lifes and single figures for close literary and artistic friends and patrons in developing Cubism.

In addition to Kahnweiler's success in displaying Cubist works at foreign exhibitions and selling them to dealers and collectors in Germany, the Netherlands, Russia, the United States and other countries, the relatively rapid diffusion and adoption of Cubist practices and ideas was considerably aided by their promotion by the self-consciously organized group of 'Salon' Cubists from 1909 to 1914, a few of whom had visited Kahnweiler's gallery and knew about the recent works of Braque and Picasso (Cox, 2000, p 136). Most of these artists lacked the connections to private dealers and collectors seeking novel art works that supported Braque and Picasso in their aesthetic project and so sought to achieve public renown through the Salon system amidst the intensified competition between artists resulting from the large increase in the number of French and foreign painters in Paris in the 1900s.

By taking over the *Commission de Placements* of the 1911 *Salon des Independents* Henri Le Fauconnier, Albert Gleizes, Jean Metzinger, Leger, Delaunay and others were able to ensure the hanging of their Cubist paintings together in the same room. Eighteen months later they created the *Maison Cubist* at the 1912 *Salon d'Automne* and, together with writing pamphlets, favourable reviews and promotional articles in the numerous little magazines circulating in Paris at the time, succeeded in generating considerable press and political attention, much of it negative, and diffusing their aesthetic practices and purposes in much of Europe and North America (Cottington, 1998, pp 40–53; 2004, pp 4–17). These exhibitions and articles contributed to 'Cubism, as a movement and as a style, (acquiring) the hegemonic status across western modernist culture that it attained by 1914' (Cottington, 2022, p 157).

The rapid expansion of the number of little magazines published in France, with more than 325 published in Paris between 1872 and 1894 and 250 literary reviews appearing between 1870 and 1914, was encouraged by: (a) the doubling of self-described writers in France between 1860 and 1890, (b) the relaxation of laws governing the French press in the early 1880s and

(c) technological improvements in newspaper publishing (Cottington, 2022, p 123). While most of these *petites revues* were literary and political rather than being concerned with painting and sculpture, critical analyses of new paintings and sculptures:

> grew in … status and in volume, as the dealer-critic network expanded, and both the criticism and the artists it addressed benefited correspondingly … moreover writing criticism about new art gave such writers a means to present an aesthetic position alongside their literary works, and to make a name for themselves. … As a result, the *petites revues* quickly came to be a significant means both of support for, and influence upon, the emergent population of innovative artists making paintings and sculptures increasingly independently of the mainstream. (Cottington, 2022, p 126)

This rapid diffusion of Cubism as a radical change in aesthetic practices and conventions was also assisted by the increasing diversity of aesthetic authority in the French – and indeed that of much of Western Europe – art world and its dominant beliefs in the early 20th century. Whereas the diffusion and legitimation of the impressionist style took around 30 years to accomplish in the relatively integrated art world dominated by the Academic/Salon system in the 1860s and 1870s, by the mid-1900s the proliferation of market niches for exotic art, diverse aesthetic positions and means of communicating them to varied audiences had considerably weakened the authority of this system (Sgourev, 2013).

In addition to the number of private dealers in contemporary art in Paris becoming sufficiently large and diversified to provide an alternative art market to the Salons by 1910, and the increasing diversity of critical commentaries on new artistic approaches, the dominant role of the state *Ecole des Beaux Arts* in educating and selecting recognized artists had become considerably reduced by the expansion of the private *academies libres* art schools from the 1870s onwards. In this period, their role changed from preparing students for the state entrance exam to supplanting its teaching and breaking its monopoly through sheer weight of numbers, becoming more open to a wider range of aesthetic ideas and practices in the process (Cottington, 2022, pp 107–8), as well as contributing to the development of the professional avant-garde in Paris (Cottington, 2022, pp 142–9).

b) Metafiction as a complementary innovation in anglophone literature

The nature of the innovation

Metafiction is a style of prose narrative that 'self-consciously draws attention to its status as an artefact in order to pose questions about the relationship between fiction and reality' (Waugh, 1984, p 2) and rejects attempts to create consistent, verifiable three-dimensional narrative worlds that were characteristic of the 19th-century realist novel (Crossthwaite, 2019, p 40). In contrast to earlier literary modernists' attacks on such novels that emphasized the role of the author in constructing coherent fictional worlds, the Anglophone metafictional novels that were published in the 1960s and 1970s focused on the constitutive role of linguistic conventions in creating fictional narratives and characters. They showed that not only is the 'author' a concept produced through previous and existing literary texts but that creating a novel is not fundamentally different from creating one's own 'reality' in that both involve the use of frames to construct and organize experience. As Patricia Waugh puts it:

> Metafiction lays bare the linguistic basis of the 'alternative worlds' constructed in literary fictions. Through continual shifts in and therefore revelations of context, metafictional texts expose the creation/description paradox. ... metafictional texts show that literary fiction can never imitate or represent the world but always imitate or represent the discourses which in turn construct that world. However, because the medium of all literary fiction is language, the 'alternative worlds' of fiction, as of all any other universe of discourse, can never be totally autonomous. (Waugh, 1984, p 100)

Additionally, the postwar metafictional writers emphasized the role of the reader in actively creating a fictional universe. As Linda Hutcheon (1980, p 27) has suggested:

> As the novelist actualizes the world of his imagination through words, so the reader – from those same words – manufactures in reverse a literary universe that is as much his creation as it is the novelist's. This near equation of the acts of reading and writing is one of the concerns that sets modern metafiction apart from previous novelistic self-consciousness.

Furthermore, metafictional authors highlighted the constitutive role of linguistic conventions in literary criticism and the consequent commonalities between writing novels and constructing critical analyses of them, questioning the ability of critical language to refer objectively and authoritatively to the literary text (Currie, 1995). In emphasizing and commenting on the discursive nature of fictional narratives, postwar metafictionalists highlighted both the linguistic constraints that ordinary languages impose on authors and readers and the possibilities available within those constraints, thus framing literary criticisms (Waugh, 1984, p 24–7).

Developing the innovation

Metafiction became a significant innovation in British and American literature in the 1960s and 1970s. In the US, early publications included Brautigan's *Trout Fishing in America* (1967), Coover's *The Universal Baseball Association, Inc., J Henry Waugh Prop.* (1968), Sukenik's *Up* (1968), Vonnegut's *Slaughterhouse-Five* (1969) and Sorrentino's *Steelwork* (1970) and *Imaginative Qualities of Actual Things* (1971) (Klinkowitz, 1980, 2017). These 'disrupters' of established literary conventions in Anglophone fiction varied in the extent to which they rejected the assumptions and procedures of the realist novel, but all self-consciously drew attention to the constructed nature of their fictions and the choices that all novelists had to make (Klinkowitz, 1980, Preface; Waugh, 1984, pp 2–19).

Raymond Federman, for example, continually rehearses possible narrative strategies in his *Double or Nothing* (1971), such as that between using the first or third person narrative to tell the story:

> FIRST PERSON
> or
> THIRD PERSON
> FIRST PERSON is more restrictive more subjective more
> personal harder
> THIRD PERSON is more objective more impersonal more
> Encompassing easier
> I could try both ways:
> I was standing on the upper deck next to a girl called Mary
> ... No Peggy
> He was standing on the upper deck next to a girl called Mary
> ... No Peggy
> (comes out the same)
>
> (Federman, 1971, p 99)

A distinctive feature of postwar American metafictional innovations was their combination of playfulness and 'exuberance that conveyed a sense of joy rather than of malice. Yes, its practitioners would sweep away the conventions of traditional fiction, though for the most part it would happen with a great sense of humor' (Klinkowitz, 2017, p 4). Additionally, building on the absurdism and black humour of the fictional legacy from the late 1950s and early 1960s, exemplified by Joseph Heller's *Catch-22* (1961), 'a strong tendency in US writers (has been) to respond to the anonymous, frenetic and mechanized society they see around them with fiction that is similarly depersonalized, hyperactive and over- or under- systematized' (Waugh, 1984, pp 115–6). Later in the 1970s, historiographic Metafiction, in which the historical record itself is represented as a fictional creation, became a distinct metafictional form. In Robert Coover's *The Public Burning* (1977), for instance, historical accounts of political events in the 1950s are treated as fictive constructs similar to the many other ones available to the author (Klinkowitz, 2017, p 7).

Many 1960s and 1970s American Metafictions attracted considerable readerships and market success beyond the academy, particularly Kurt Vonnegut's 1969 *Slaughterhouse-Five* (Klinkowitz, 1980, pp 33–7). Later works that adopted metafictional techniques more wholeheartedly and engaged in what Patricia Waugh terms 'intertextual overkill, in which … texts/writing is explicitly seen to produce texts/writing. Linguistic codifications break down into further linguistic codifications' (Waugh, 1984, p 145), however, tended to appeal to a more limited segment of critics and readers, mostly academic in nature.

Nevertheless, metafictional techniques:

> have become less of a confrontational challenge to tradition than a helpful supplement to it … metafiction has been accepted less as a destructive tendency that must overthrow mainstream fiction (or cease to exist itself) and more as a corrective to what had become a monological tendency among novelists to write fiction in only one way. (Klinkowitz, 2017, p 13)

Conditions supporting the development of the innovation

In addition to the 'expansionary publishing conditions of the 1960s and early 1970s (that) were relatively hospitable to authors experimenting with alternative narrative forms' (Crosthwaite, 2019, p 40), following the paperback revolution that greatly enlarged the market for novels (Stevenson, 2004, pp 137–43), the rapid development of early metafictional novels in the UK and USA was assisted by two sets of changes in the socio-economic environment of literary production and readership. First, the cohesion and aesthetic authority of the elite cultural networks that had connected academies, museums, art and music associations and learned societies with various literary establishments declined considerably as the 'Humanities Revolution' resulting from the rapid and large scale expansion of universities and colleges took place between 1945 and 1975 (Menand, 2010, pp 63–84). As Leypoldt (2015, p 769) suggests:

> A crucial aspect of this shift concerns the conditions of authorship: the growth of third-level education produced new systems of literary patronage that provide non-commercial writers with university-embedded social-professional networks which connect creative writing

programs and English departments with editors and publishers in a transatlantic print market.

These new networks integrated the production of new novels with academic analyses and blurred the boundaries between critics, teachers and writers. Particularly with the post 1965 growth of creative writing programmes in US universities that combined the employment of practicing authors with the recruitment, education and certification of would-be writers, usually located within established departments of English, they provided economic and cultural support for innovative approaches in both literary production and critical practices (Myers, 2006; McGurl, 2009).

The development of successful academic careers by creating novel critical positions and techniques in an expanding higher education system, often invoking influential deconstructionist arguments, became combined with the construction of influential writing careers drawing upon these approaches, as in the cases of Ronald Sukenick and Raymond Federman who completed their PhDs at leading institutions before publishing their first novels. As Klinkowitz (2017, p 8) emphasizes, in an age when serious fiction was increasingly nurtured by the universities these two high profile academics were able to exert an innovative pressure on literary history well in excess of their own novels' marketing success.

Second, the combination of the postwar expansion of higher education in the humanities and the growth of creative writing programmes in universities greatly enlarged the number of people able and willing to invest time and effort in reading and understanding specialized and difficult kinds of fiction. This was particularly so for works that attacked established practices with humour and emphasized their artificiality and dependence on questionable assumptions about the nature of language and descriptive accounts of everyday actors and situations (Waugh, 1984). Although some of the established creative writing programmes had traditionally focused on the production of realist novels, their postwar expansion in popularity among writers and universities enabled experimental novelists to considerably enlarge the audience for their iconoclastic approaches and practices (McGurl, 2009).

c) The Oncogene Approach as a displacement innovation in cancer research

The nature of the innovation

This displacement innovation occurred in the late 1970s and early 1980s when the combination of the proto-oncogene theory of the causation of cancer and the adoption of molecular-biological research methods enabled the problems posed by the new theory to be studied empirically replaced three largely unconnected approaches to cancer research.

Until the mid-1970s, there were three distinct approaches to the study of cancer:

- The heritability of cancer was investigated with strains of inbred mice which were assumed to be genetically identical. Crossing mouse strains that produced tumours with strains that had a propensity not to produce tumours was expected to establish the heritability of cancer and the role of spontaneous mutations (Fujimura, 1996, p 26; van Helvoort, 1999).
- Research on exogenous causes of cancer used mice and cell cultures to establish the causation of cancer by chemical compounds, hormones and radiation (Fujimura, 1996, pp 53–64). This research was primarily theoretical and biochemically descriptive until the late 1970s (Fujimura, 1996, p 63).
- The discovery of the mouse mammary tumour virus was an impetus and justification for the search for tumour-causing viruses from the 1960s onwards in the major US government-funded research programme on viral carcinogenesis. Viruses became the most popular hypothesized cause of cancer, and genetics moved into the background (Fujimura, 1996, p 36).

These approaches lacked both a common theoretical understanding of the causation of cancer and the tools to investigate the questions they posed empirically. According to Fujimura:

> until the mid-1970s, researchers had no technologies for experimenting with theories of cancer in eukaryotic organisms at the molecular level. Until 1976 basic cancer research consisted of many theoretically and organizationally different lines, most of which were far from an empirical description of the molecular genetic mechanisms of cell transformation. The molecular biology of cancer was limited to studies of tumor viruses and viral oncogenes and the protein chemistry of various proteins detected in transformed cells. (Fujimura, 1996, p 64)

The dominant assumption at the time was that there is 'a "factor" existing within the cell that is critically involved in producing tumor' (Fujimura, 1996, p 64). However, cancer researchers had no means of identifying this factor.

This set of disparate assumptions about the cause of cancer was replaced by the Oncogene Approach that combined the 'oncogene theory' – the assumption 'that cancer is caused by normal cellular genes called "proto-oncogenes" that somehow turn into cancer genes, loosely called "oncogenes"' (Fujimura, 1988, p 262; see also Morange, 1997, p 7) – with recently developed molecular biological methods to investigate these genes. The oncogene theory provided the common theoretical background to the previously weakly connected strands of cancer research by proposing a mechanism of cell transformation that could be triggered by each of the causes under investigation (van Helvoort, 1999). A major factor promoting the transition to the new theory was the parallel development of the process innovation of molecular-biological methods that allowed the precision manipulation of genetic material in eukaryotic cells. In contrast to previous cancer research, the questions derived from the new theory could thus be empirically investigated.

Developing the innovation

The oncogene theory developed from two major sources. First, the search for viruses causing cancer in humans, while largely unsuccessful, led to the discovery of proto-oncogenes, that is genes in normal human cells that could be transformed into cancer-causing oncogenes. This discovery was made after researchers found specific cancer genes in viruses that caused tumours in cultured cells and laboratory animals (but not in wild animals or humans). Subsequent research found genes that were very similar to those cancer genes in normal cells of many primates. In the late 1970s, this evidence was reinterpreted as viral oncogenes coming from normal cellular genes rather than vice versa (Fujimura, 1996, pp 117–9).

This hypothesis was independently confirmed by molecular-biological DNA transfection studies in the beginning of the 1980s. The transfer of DNA from tumour cells to healthy cells turned them into tumour cells. These experiments confirmed that the unknown 'factor' that produces tumours is in fact the genetic material of the cell. When researchers identified the genes that transferred the information for being a tumour cell, they learned that these were the same oncogenes identified by virus researchers (Fujimura, 1996, pp 123–7).

The oncogene theory provided researchers with both a mechanism to study the transformation of the DNA from healthy cells into tumour-producing DNA and concrete empirical objects to study and to search for (oncogenes and proto-oncogenes). This empirical research was made possible by the development of new recombinant DNA technologies that enabled the manipulation of eukaryotic cell DNA. Before these technologies were developed, researchers had to focus on the investigation of prokaryotes – simple organisms like bacteria, viruses or algae whose cells have no defined nuclei (Fujimura, 1996, p 73). Key contributions to the development of recombinant DNA technologies were: first, the discovery of restriction enzymes that cleave DNA at highly specific points in the molecular chain in bacteria, second, improved methods for sequencing DNA and, third, improved methods for synthesizing short strands of DNA (so-called oligonucleotides). These methods made it possible to 'cut and paste' DNA fragments and to recombine DNA from different species.

Recombinant DNA technologies profoundly changed empirical research in many fields of the biomedical sciences and beyond. However, while there was little opposition to the new genetic approach to explaining cancer, turning to the new approach required a significant investment in learning the new molecular biological techniques. Initially, these recently developed methods for investigating genetic causes for cancer and mechanisms that turned normal cell into cancer cells were weakly codified and difficult to learn. Their early circulation was limited and depended on laboratory courses and researcher mobility. To become widely used, they needed to be standardized so that they could be easily learned and transmitted. This standardization occurred in the late 1970s and early 1980s and included the use of 'pre-fabricated biological materials (reagents, probes), procedural manuals (often called "cookbooks") spelling out "recipes" for action, industrial standards, computing protocols, and instruments that automate many procedures' (Fujimura, 1988, p 266).

These new methods facilitated development in all strands of cancer research. The DNA transfection studies mentioned previously already utilized molecular-biological methods. For the study of carcinogens, Fujimura quoted a researcher who suggested that: 'Despite our great ignorance about the basic biology of cancer, the study of carcinogens has recently been made much easier by the application of knowledge and techniques derived from bacterial molecular genetics' (Cairns, 1978, p 90, quoted after Fujimura, 1996, p 63). More generally, recombinant DNA technologies made possible the study of a wide range of problems linked to human cancers which attracted many researchers at universities as well as private companies. Researchers could turn to these new practices without having to forsake previous commitments to other theories because the hereditary, environmental and viral causation of cancer could still be investigated.

Conditions supporting the development of the innovation

The Oncogene Approach was developed in a practitioner-led science world in which the scientific community shared authority with two kinds of intermediaries, namely science policy makers and administrators who were interested, and significantly invested, in the search for a cure for cancer and the pharmaceutical industry which had a commercial interest in developing cancer therapies. The administrator roles were often occupied by practitioners from scientific communities engaged with or close to cancer research.

A major condition facilitating the development of the innovation was the strong political commitment in the US to finding a cure for cancer. The

National Cancer Institute (NCI) – a division of the National Institutes of Health – had funded research on cancer viruses since 1964 through its Viral Cancer Program. In 1970, President Nixon declared 'War on Cancer'. The subsequent substantial funding for cancer research under the National Cancer Plan of 1971 focused on the study of 'Viruses and Cancer' because it was expected that this line of research would lead to fundamental breakthroughs. In this Plan, 'the Leukemia Program was converted to the Special Virus-Cancer Program, which was designed to search systematically for possible viral agents of cancer in man' (van Helvoort, 1999, p 318), despite a lack of evidence that viruses produced human cancers, a move that 'drew heavy criticism from biological and biomedical scientists' (Studer and Chubin, 1980, pp 97–8; Fujimura, 1996, p 131).

As well as this concentration of funding on one of the three major lines of cancer research to the detriment of alternative lines of inquiry being criticized by the scientific community, there was also concern about the possible overfunding of the field. Fujimura quotes one of her interviewees saying that this programme 'probably sponsored more crappy research than almost any scientific campaign I can remember' (Fujimura, 1996, p 132). In addition to its thematic focus, the programme was also attacked for its large share of funding being allocated for contractual research with private enterprises and organizations and for its inequitable distribution of funds (Studer and Chubin, 1980, pp 95, 97–8; Fujimura, 1996, p 131).

These criticisms made the then-director of the NCI Rauscher establish an ad-hoc committee led by Dr Norton Zinder from Rockefeller University to review the Special Viral Cancer Program (VCP). It submitted an extremely critical report to the National Cancer Advisory Board (NCAB), which oversaw the work of the entire NCI (Studer and Chubin, 1980, pp 83–7; Fujimura, 1996, p 131). The report was particularly critical of the unequal distribution of funding (most of the funding went to very few researchers) and of the large proportion of contract funding (Studer and Chubin, 1980, pp 84–5, 95–6).

In 1980, NCI leaders decided to break up the VCP and integrate the pieces with other NCI programmes (Fujimura, 1996, p 132). Since the NCI-funded research on viruses had led to the discovery of oncogenes and proto-oncogenes, the SVCP was retrospectively justified with these findings even though it failed its main purpose of identifying viruses that cause cancer. The oncogene theory was presented as the outcome of the search for cancer viruses and as the new focus of funding for cancer research.

Funding was refocused on molecular-biological research derived from the oncogene theory (Fujimura, 1996, pp 130–5). This reorientation was widely accepted because the search for cancer viruses was considered to have exhausted itself, because the new oncogene theory was presented as the outcome of previous funding and because the oncogene theory was used to communicate a new urgency for finding a cure. One NCI administrator used the oncogene theory in communications with Congress by pointing out that all members of Congress had proto-oncogenes in their genetic material:

> The audience listening to DeVita's discussion of oncogenes understood these genes to be the first step in the process toward tumor and death. However much their views may differ in terms of 'scientific' nuance or sophistication, the statement that we all have genes that can be triggered to cause cancer can engender great fear in congressional members and their constituents because they share the basic notion of genetic causality. This fear, however, was laid to rest by the claim that a unifying pathway to all cancers may exist and that there may be ways to intervene in this pathway. Thus, DeVita found 'oncogenes' useful for translating the interests of the sitting members of Congress. (Fujimura, 1996, p 134)

The new funding for oncogene research was accompanied by rapidly growing public and private investments in the development of molecular-biological methods (Fujimura, 1996, pp 88–100). In addition to rapidly rising NIH funding for molecular-biological research, private industry invested substantially. Small specialized companies provided reagents and instruments for molecular-biological methods, specialized tests and software. This exploitation of commercial opportunities contributed to the quick standardization of the new molecular-biological methods in the late 1970s and early 1980s and thus to the rapid development of the Oncogene Approach. Larger corporations focused their in-house research on product development but also funded basic research at universities, which they hoped might lead to new products. According to Fujimura, '[l]arge corporations invested a total of about $250 million into biotechnological research contracts with universities and research institutes in 1981 and 1982' (Fujimura, 1996, p 88, see also Fujimura, 1996, p 90). Although these investments contributed to the rapid development of the Oncogene Approach by attracting researchers and providing research opportunities, the development of the innovation was largely directed by researchers responding to scientific opportunities and predominantly funded by the state through the NIH.

The oncogene theory and molecular-biological research remained the dominant approach to investigating cancer. Their continuous development is made possible not only by their continued provision of opportunities for researchers to produce relevant contributions to cancer research but also by heavy investments by governments and private industry, which in turn responded to the perception in society of cancer as one of the major health risks.

d) The rise of Financial Economics as a displacement innovation in the social sciences

The nature of the innovation

The transformation of business finance research and teaching in leading North American business schools into Financial Economics between the 1960s and the 1980s represented a major change in both the nature of research practices and problems and the distribution of epistemic and resource allocation authority. Prior to the 1960s, much of the research literature in the business finance field as represented in the dominant journal – the *Journal of Finance* – covered a variety of financial topics, was primarily descriptive and not theoretically integrated (Kavesh, 1970). Summarized as the 'traditional trinity of money and capital markets, business finance, and investments' (Weston, 1974, p 239), common subjects included the nature and functioning of financial institutions, corporate financial management and the long-term financing of companies and investment projects, with relatively little attention being paid to the theory and operation of capital markets and management of risk.

Research results did not require a specialized academic training to be understood and often addressed problems that were not sharply distinguished from the concerns of senior finance practitioners and advisors, some of whom contributed to journals and served with academics as officers of the American Finance Association (AFA). Overall, business finance research did not appear to be tightly controlled by academics nor were research skills highly standardized or formally institutionalized in doctoral programmes. Most courses in postgraduate business schools were concerned with technical procedures and practices for corporations raising long term finance, investment practices and the operation of the banking system rather than theoretically derived models (Whitley, 1986a; MacKenzie, 2003; Fourcade and Khuran, 2013).

By the 1980s this situation had changed considerably. The active membership of the AFA had more than doubled to over 3000 and almost all of its officers and directors were academics. Practically all the papers published in the *Journal of Finance* and the other two major journals which had been founded since 1960 – the *Journal of Financial and Quantitative Analysis* and the *Journal of Financial Economics* – relied extensively on mathematical and/or statistical analysis and focused on asset valuation models and capital market price movements in highly idealized markets. The central questions of the field now concerned the efficiency of financial markets rather than the funding of firms. As well as research techniques and the language for communicating research becoming much more abstract, formal and restricted to professional researchers, the topics and problems studied in finance research shifted greatly in this period.

In 1964, 27 of the 32 papers accepted for publication in the *Journal of Finance* were classified under two broad and diffuse headings: Business Finance and Investments and Money, Banking and Financial Institutions. In 1980, there were seven categories for classifying contributions, of which only one was the same as those used in 1964. Twenty-two of the 72 papers accepted were classified as theoretical while 19 were characterized as empirical studies of financial institutions and markets. These latter papers were mostly statistical analyses that had little in common with those published on overtly similar topics in the early 1960s. Essentially, research in business finance had been transformed into a branch of neo-classical micro-economics concentrated on highly abstract models of stock price movements, partly based on empirical studies of successive stock and commodity price movements that demonstrated the lack of serial correlations between such movements in any one period and those in the succeeding one (Kendall, 1953; Keenan, 1970; Weston, 1974).

Similar changes had occurred in the structure and content of finance teaching at the major business schools. New textbooks emphasizing the analytical derivation of financial management models and techniques from micro-economic choice theory have become dominant in many schools, and the teaching of corporate finance was increasingly seen as part of a general introduction to the theory of asset valuation in perfectly competitive capital markets.

Developing the innovation

These changes reflected the widespread adoption of neo-classical economists' assumptions about the nature of perfect markets in equilibrium – that is, when supply and demand of financial assets are equal so that markets clear – and the subsequent investigation of their implications for the pricing of such assets under particular conditions. Four approaches in particular generated a large number of theoretical and statistical elaborations and further development: the efficient (stock) market hypothesis (EMH), the irrelevance of firms' capital structures for their market valuation, the Modern Portfolio Model for financial assets and the Capital Asset Pricing Model.

The efficient market hypothesis suggested that the lack of serial corelations between movements of stock and commodity prices in any one period and those in the succeeding one, and the consequent adoption of a random walk model as the best model of price changes, was due to the efficiency of market actors in using relevant information to price claims on future income. Essentially, the EMH states that in determining the equilibrium price of a financial asset all the relevant and widely available information about it has been used and is reflected in the current price. Necessarily, then, any subsequent change in the price of that share will reflect new information only. Such new information is, by definition, unpredictable and so too is the next change in its price, ensuring that efficient markets must generate random movements in prices. While Irwin Friend (1973), in his presidential address to the AFA, pointed out that 'this conception does not consider the relevance of the information to the subsequent earnings or riskiness of the stock and does not distinguish between a market in which information is sparse and of low quality and one in which it is abundant and of high quality' (Friend, 1973, p 269), this idea became widely adopted in the 1970s (Shiller, 2003).

The EMH was assumed to be correct in two influential papers by Modigliani and Miller (1958) and Miller and Modigliani (1961) about firms' capital structures, dividend policy and valuations that led to their work being awarded the Nobel prize in economics. In these contributions, they assumed the existence of perfect capital markets in which all traders are price takers, have equal and costless access to information, are not subject to brokerage fees or other transaction costs or differential taxation, and investors are 'rational' in being indifferent about taking returns through dividends or capital gains. On this basis, they concluded that firms had no rational basis for preferring debt to equity finance and dividend payments are irrelevant to asset valuations.

Despite some claims that empirical evidence supports this model, Friend has emphasized that:

> The obvious fact is that under the MM theory, so long as the rate of corporate taxation is substantial and the difference between the effective rates of personal taxation of income from corporate bonds and stock is relatively small, the minimal cost of capital is achieved at a capital structure which consists entirely of debt. Since the average post-World War II ratio of debt in the corporate capital structure in the U.S. seems to have been less than 30%, it is difficult to take seriously any empirical verification of the MM theory. (Friend, 1977, p 115)

Nonetheless, it has become regarded as a significant contribution to the development of Financial Economics and stimulated numerous theoretical modifications and elaborations.

A separate strand of work involved in the transformation of business finance concerned models of rational investor behaviour in constructing investment portfolios. Later summarized as Modern Portfolio Theory (MPT), this was initially outlined in an early paper by Harry Markowitz (1952) who suggested that investors should only be concerned with two factors: the expected return from holding a particular stock and the expected risk involved in holding it. Risk here refers to the expected variance of returns over the period for which the investment is held rather than the risk of holding equity assets in general. Rational behaviour is here assumed to mean that investors seek to maximize expected returns while minimizing the volatility of those returns.

Since the combination of any two investments with different expected variances of returns, and whose returns are not perfectly correlated, reduces the overall level of such risk, investors can minimize their exposure to risk by diversifying their holdings across investments whose expected returns and variances differ considerably. While such diversification does not reduce the overall risk of holding shares in general, or 'market' risk, investors can choose whether to construct portfolios that exaggerate or mitigate market risk by focussing on stocks that are more or less sensitive to market movements as measured by the covariance of expected returns of holding a stock with those of a total market portfolio.

Together with the assumptions underpinning the EMH, Markovitz's work, later elaborated in his 1959 book *Portfolio Selection*, was used in subsequent studies of portfolio risk premiums and in the development of the Capital Asset Pricing Model (CAPM) (Sharpe, 1964; Lintner, 1965). This model became the focus of a great deal of theoretical and empirical attention – to the extent of being termed the 'paradigm' of Financial Economics (Ross, 1978; Ryan, 1982; Mangee, 2015), and led to Sharpe being awarded the Nobel prize in economics in 1990.

It asserts that the price of a claim on future financial returns depends on its risk, understood as volatility, relative to the market portfolio. The greater is the contribution of an asset's variability of returns to the overall variability of a market portfolio, the higher has the expected return to be for an equilibrium price to be derived. Given the seven basic assumptions of the model, as summarized by Jensen (1972), the price of an asset at equilibrium is given by a function of the risk-free interest rate, the mean expected returns generated by holding that asset and the covariance of those expected returns with the returns on all assets in the market portfolio. Numerous attempts to 'test' this model have usually presumed that actual returns reflect expected returns and so should be positively correlated with their covariance with indices of market returns, producing rather inconclusive results (Friend et al, 1978).

These theorems and models were elaborated and 'tested' against econometric studies of price movements in stock markets over the next 20 or so years. Much of the theoretical modelling involved the application of novel mathematical techniques and minor modification of assumptions to account for apparent deviations from random walk models of successive price changes and other theoretically irrational patterns of behaviour in financial markets. The 1960s and 1970s also witnessed a rapid expansion in the number of PhD theses applying econometric techniques to the empirical analysis of stock and bond price movements with the aid of computers and ready access to contemporary and historical stock market records (Fourcade and Khurana, 2013).

Some of the problems raised by these studies stimulated the creation of an Arbitrage Theory of Capital Asset Pricing by Stephen Ross to replace the CAPM in the 1970s and in the 1980s behavioural finance emerged as an alternative approach that emphasized the bounded rationality of decision makers and the limits to arbitrage (Ross, 1976; Friend, 1977; Shiller, 2003; Jovanovic, 2008; Mangee, 2015). In the 1990s and 2000s the persistence of anomalies in capital market behaviour encouraged further alternative approaches to develop including neurofinance, evolutionary finance, the adaptive market hypothesis and sociological finance (Gippel, 2012).

Conditions supporting the development of the innovation

This transformation of business finance research and teaching into Financial Economics was facilitated by three major changes in practitioner communities and their societal contexts. First, the substantial changes in the US higher education system that occurred since the end of the Second World War, particularly the expansion resulting from the GI Bill of 1944 and the National Defense Education Act of 1958. Second, the increasing ease of access to computers and historical information about stock market price movements, dividends and returns and third, changes to the organization of capital markets and corporate finance functions that expanded the demand for analytical skills in financial markets.

A significant component of the postwar growth of US universities was the rapid growth of business education, especially at the graduate level. As Fourcade and Khurana (2013, p 122) emphasize: 'Between 1960 and 1980, MBA education grew at an annual rate of 12%. More than 55,000 MBAs were granted in 1981, surpassing the combined total of law and medical school degrees.' This expansion was accompanied by a widespread belief in the efficacy of what became known as management science techniques and models in dealing with complex organizational problems and business decision-making and the coterminous need to upgrade business school faculty and teaching by investing in scientific research, as exemplified by the Carnegie and Ford Foundation sponsored reports on high level business education (Gordon and Howell, 1959; Pierson et al, 1959). The Ford Foundation, in particular, played a major role in supporting business school PhD programmes and the recruitment of discipline trained faculty from mathematics, economics and statistics, especially at the Graduate School of Industrial Administration at Carnegie Tech, and the business schools at Chicago, Harvard, Stanford and Columbia universities (Fourcade and Khurana, 2013, p 143).

The rapid growth of empirical research on equity asset valuations and corporate capital structures in the 1960s and 1970s was greatly facilitated by the growing availability of large computers and data bases on stock market price changes, which enabled business schools to generate a large number of PhD theses and publications on stock market price changes. The Chicago Graduate School of Business benefited considerably from the series of grants from investment bank Merrill Lynch for their Center for Research in Security Prices, which generated a considerable growth in the number of PhD theses on price movements in capital markets (Fourcade and Khurana, 2013, p 148). These data bases and allied technological improvements formed the basis of an econometric research community focusing on the analysis of market changes (Whitley, 1986b).

The growth of Financial Economics was additionally encouraged by major shifts in the organization and control of financial markets. These included the expansion of collective investment organizations, such as pension funds and unit trusts, their growing regulation through such legislation as the Employee Retirement Security Income Act, which focused attention on the measurement of risk and the development of risk management techniques based on Modern Portfolio Theory, and the intensification of competition between financial intermediaries in the 1970s. As a result, the demand for academically educated and certified financial analysts grew considerably, increasing the willingness of university business schools to invest in finance research and teaching.

These shifts led to a variety of financial innovations being developed, especially the development of new financial instruments and markets such as options, futures and credit default swaps (MacKenzie, 2003; MacKenzie and Millo, 2003), memorably described by Warren Buffet in his 2002 Chairman's letter to Berkshire Hathaway shareholders as 'weapons of mass financial destruction'. This unusually rapid and far-reaching displacement innovation in the social sciences was supported, then, from the combination of substantial expansion of training and employment provision in higher education, technological changes and major shifts in the structure and size of institutions and capital and labour markets in the financial services sector.

Summary

In these four case studies of major innovations in practitioner-led art and science worlds, both practitioner communities and intermediaries changed roles in ways that led to the redistribution of authority between different intermediaries and enabled new practices and approaches to dealing with constitutive problems to be developed and adopted quite quickly. As with Bebop and the *Nouvelle Vague* these innovations were additionally supported by significant changes in the concerns and interests of their audiences, both those directly involved in the recognition and assessment of novelties and those investing in their distribution and consumption.

In the case of Cubism, the dominant role of the state Academy and Salon exhibition system in the training, certifying and recognition of artists and their works had been already considerably weakened by the growing aesthetic and commercial success of the Impressionist and Post-impressionist movements over the previous three decades as a result of major changes in the art world that enhanced the authority of practitioners. These included the expansion of the private *academies libres* and increasing variety of Salon exhibitions, the emergence of a range of dealers varying in size, stability and commitment to supporting innovative styles and approaches to artistic problems, and the diversity and internationalization of art markets across Europe and the Atlantic.

In the case of Metafiction, following the postwar expansion of the novel reading public, similar shifts in the literary art world enabled experimental writers to explore the role of linguistic conventions in constructing both fictional worlds and everyday experiences. Additionally, the growth of English departments in leading universities increased the authority of academic intermediaries and writer-critics over aesthetic goals and standards at the expense of established literary critics as well as providing new employment opportunities for both critics and authors. The related growth of creative writing programmes at many universities also generated a substantial group of readers and commentators who were knowledgeable about, and interested in, the practices of writing fictional narratives. More generally, this expansion increased the wider market for unusual and self-consciously deviant contributions at a time of political turmoil and cultural conflicts.

In the case of the Oncogene Approach, practitioners acted as science policy makers and administrators in refocusing the public funding of cancer research on the molecular-biological analysis of oncogene development and searching for ways of intervening in these processes. The combination of substantial private investment by large and small firms with this public support for the development and standardization of molecular-biological methods, tests and software for oncogene research encouraged scientists focused on exogenous causes of cancer to accept and develop the Oncogene Approach, accelerating its adoption by the practitioner community as a whole.

The case study of Financial Economics highlights the changing roles of intermediaries in supporting the development of Financial Economics as a displacement innovation. In addition to the postwar expansion of higher education and federal support for scientific research in universities encouraging administrators to invest in the new behavioural sciences, the Ford Foundation and similar organizations pressured business schools to improve their research activities considerably and funded a number of PhD programmes in what was considered the more scientific areas. Together with corporate and personal support for technological facilities and data sets as well as for free market beliefs in the efficiency of asset markets, these changes enabled finance researchers to integrate their field with orthodox economics and tighten academic control over epistemic standards and goals.

Innovation development in practitioner-dominated science worlds

a) Radio Astronomy as a process innovation in the physical sciences

The nature of the innovation

Radio Astronomy was a process innovation that solved three major problems of Astronomy. At the beginning of the 1940s, astronomers knew that the largest part of the Milky Way including its core was inaccessible to optical astronomy due to the extinction of light by interstellar dust, which meant that many questions about the Milky Way and other galaxies could not be answered. Second, optical astronomy in many regions including Northern Europe was constrained by the cloudy climate, which severely limited the time for astronomical observations. Third, astronomical observatories close to growing cities faced the problem of increasing light pollution, which reduced the quality of observations (Woerden and Strom, 2006; Elbers, 2012; Munns, 2013).

The utilization of cosmic radiation – radiation that does not originate from earth – for astronomical observations was triggered by its initial discovery in telephony. Cosmic radiation was first discovered in 1937 by Jansky, an employee of the Bell Telephone Company, who was tasked with identifying the sources of the static that interfered with radio-telephones. He identified the Milky Way as one of the sources of this static. His publication motivated a radio engineer, Reber, to build dedicated equipment for mapping the celestial distribution of cosmic static. Reber published his observations in 1940 in the *Astrophysical Journal* and contacted several US universities with plans to build a radio astronomy observatory, all of which turned him down (Mulkay and Edge, 1973, pp 25–6; Elbers, 2012, pp 266–7; Munns, 2013, p 38). Nevertheless, optical astronomers noticed the discovery and the possibility of using cosmic radiation. The development of radar techniques during the Second World War increased attention to cosmic radiation as a major source of noise interfering with radar signals and created a first generation of equipment for the detection and measurement of cosmic radiation.

The discovery of cosmic radiation provided the opportunity for scientists to utilize radiation at other than optical wavelengths for astronomical observations, that is, radiation that was not degraded by interstellar dust, clouds or obscured by terrestrial light sources. The development of new techniques and practices to study it enabled researchers to investigate: (a) areas of the cosmos that were inaccessible before, (b) astronomical objects that did not emit light in the visible spectrum and (c) a wider range of properties of known astronomical objects. Within three decades, optical astronomy evolved into an all-wavelength astronomy with a significantly broadened empirical domain.

Developing the innovation

The innovation was developed from these initial discoveries both inside and outside the astronomical community in the 1940s and 1950s. In the Netherlands, and later in the USA, Radio Astronomy was developed by astronomers who recognized its potential to overcome limitations of optical astronomy, in particular its degradation by stellar dust, which prevented optical astronomers from observing even the core of our own galaxy, and the interference of the increasing number of terrestrial light sources with optical observations.

Dutch astronomers began to search for a suitable wavelength for Radio Astronomy and to design a dedicated radio telescope in the early 1940s, during the German occupation. In 1944, a Dutch physicist identified the 21 cm spectral line of hydrogen as the potentially most useful wavelength for radio-astronomical observations. Plans to build a radio telescope were developed in close collaboration with engineers and researchers from Philips Electronics but had to be delayed due to the German oversight of Philips (Woerden and Strom, 2006; Elbers, 2012, pp 266–9). After the war, Dutch astronomers initially used radar equipment left behind by the German army for their research before the first dedicated radio telescope was built in 1955. The research focus of Dutch astronomers was galactic structure and rotation. First findings included the distribution of hydrogen in the galactic plane, data that suggested the existence of spiral arms of the galaxy, and estimates of the rotation speed of the galaxy (Woerden and Strom, 2006).

In Australia and the UK, Radio Astronomy was first developed by radar physicists who constructed and applied ever-more sensitive radar equipment during the Second World War. Already used for military purposes during the war, the antennas of this equipment had to be turned to the sky occasionally to identify background noise. In 1945, New Zealandian and Australian radio physicists used it to identify sun spots as a source of radio emissions, thereby showing that the detection of extraterrestrial radio emissions was possible (Munns, 2013, pp 27–8). The Australian group conducting radio astronomy, the Radio Physics Laboratory, was part of the Commonwealth Scientific and Industrial Research Organisation (CSIRO), a state-funded research organization that was mainly devoted to applied research. The Radio Physics Laboratory was founded during the Second World War to assist the Australian Defence forces with the development of radar technologies. After the war, the group conducting solar noise research expanded as alternative projects for radar physicists failed (Munns, 2013, pp 66–9). This group detected the first extragalactic radio source (Cygnus A) and the first correlation between a radio source and an optical object (the Crab Nebula) in 1948 (Munns, 2013, p 69). These discoveries were recognized by optical astronomers in the Netherlands and the US, strengthening their interest in Radio Astronomy and leading to international collaborations (Munns, 2013, pp 69–71).

In the UK, groups of radar physicists were established at the University of Manchester by Bernard Lovell and at Cambridge University by Martin Ryle. British radar physicists did not have clear radio-astronomical research programmes in the immediate postwar years (Munns, 2013, p 74). Lovell's group began its work at the end of 1945 with the application of radar to the detection of meteor showers. Using radar equipment which they easily obtained from the military after the war, they were able to resolve a controversy between meteor astronomers by establishing that sporadic meteors were members of the solar system. The investigation of meteors led to the first contacts between the radar physicists, who had to familiarize themselves with the astronomical problem, and meteor astronomers, who were invited to Jodrell Bank, the location of the university of Manchester's radar physics group (Gilbert, 1976, pp 187–90).

At the end of the 1940s, both Lovell's and Ryle's groups became more concerned with the astronomical significance of their research. In July 1949, Lovell hosted 60 Fellows of the Royal Astronomical Society in Manchester and at Jodrell Bank in Cheshire in a two-day meeting largely devoted to the instruments at Jodrell Bank and the astronomically interesting results they had produced.

The University of Manchester group conducted observations at Jodrell Bank with instruments that included a 218-foot diameter parabolic dish that was completed in 1950 (Munns, 2013, pp 74–5), followed by a 250-foot 'radio telescope' in 1957. This equipment was used to map the Andromeda Nebula in 1950 (Munns, 2013, pp 74–5), which was the beginning of a broad survey project that triggered a collaboration with an optical astronomer in Manchester. In 1950, the Cambridge group conducted a survey of radio objects and identified 50 radio sources. From 1951 onwards, they collaborated with optical astronomers in the US.

The main problem hindering the integration of radio-astronomical with optical observations in the first years of Radio Astronomy was the latter's insufficient resolution, which was responsible for large margins of error in the positioning of radio sources and thus for their identification with objects observed by optical astronomers. This initial lack of precision of the new methods made some optical astronomers reluctant to accept the new methods as potentially useful for astronomy. The precision of radio-astronomical observations improved significantly during the 1950s, when military radar dishes, antennas and receivers were gradually replaced by dedicated radio-astronomical instruments (Edge and Mulkay, 1976, pp 134–52).

Unlike Australia and the UK but similar to the Netherlands, the first advocates for Radio Astronomy in the USA were optical astronomers. Harvard optical astronomers became concerned with increasing light pollution from 1950 onwards, which led to the decision to develop a Radio Astronomy programme in 1952, the operation of the first Harvard radio telescope in 1953 and the first PhD in radio astronomy granted in 1954. From the mid-1950s, radio telescopes were considered by the international astronomical community as astronomical instruments. In 1958, the International Astronomical Union changed its approach to galactic coordinates and the definition of the galactic pole in response to radioastronomical findings (van Woerden, 2006, p 12). From then on, Radio Astronomy can be considered to be fully integrated into astronomy, which became transformed from optical astronomy into an all-wavelength astronomy.

Conditions supporting the development of the innovation

Soon after cosmic radiation was discovered, it became obvious to astronomers that Radio Astronomy might be able to solve important problems of their discipline like the extinction of light by interstellar dust, the degradation of astronomical observations by terrestrial light sources or the dependence of optical astronomy on the weather. The three major conditions that supported the rapid acceptance of the new methods, which began in 1945 and can be said to have been achieved in 1958, included the access to (funding for) equipment, the participation of optical astronomers in the development of the innovation, and the recruitment of graduate students. All three differed between the four countries in which Radio Astronomy was first developed, that is, the Netherlands, the US, Australia and the UK.

In the Netherlands and the US, optical astronomers from the beginning developed Radio Astronomy as an innovation in astronomy because it overcame limitations of optical observations. Decisions to build dedicated radio telescopes were made by optical astronomers at the University of Leiden, Harvard University and the California Institute of Technology. These optical astronomers needed to recruit physicists and engineers to build and operate the equipment. In the Netherlands, radar equipment left behind by the German army was first used but plans to build a dedicated radio telescope were developed as early as 1944. After the war, astronomers lobbied for funding from the Dutch government which was not initially granted due to the large sums that were required. Small-scale funding for radio-astronomical research was, however, granted by the Dutch Organisation for Pure Scientific Research (ZWO). ZWO rejected the application for the telescope in 1949, partly due to the slow research progress that was made with the existing radar equipment. Funding was eventually granted in 1951 after Dutch researchers contributed to the detection of the 21 cm hydrogen line (Elbers, 2012, pp 279–80).

In the US, astronomers at Harvard University responded to the increasing light pollution by searching for funding for better equipment and by developing a collaboration with South Africa (which was rescinded in 1953 over the issue of Apartheid). They also turned to Radio Astronomy, which was rapidly integrated in the Harvard astronomy programme. In 1952, the Harvard observatory council favoured Radio Astronomy. The first radio telescope was built in 1953, and its operation was secured by a gift from a patron in 1954. The integration of Radio Astronomy into Harvard's astronomy programme was also a response to the difficulties in attracting graduate students. Graduate programmes in nuclear physics were much more popular at the time, and astronomers (correctly) assumed that offering research in radio astronomy, which was based on radio physics, would increase the attractiveness of their graduate programme (Munns, 2013).

In Australia and the UK, radio physicists originally developed their own approaches to Radio Astronomy but soon began to collaborate with optical astronomers. Australian radio astronomers responded to the antagonism of a powerful optical astronomer by developing close collaborations with optical astronomers in the US and the Netherlands, which rapidly developed after the first identification of a radio source with an optical object – the Crab Nebula (Munns, 2013, p 69). The Cambridge group in the UK also collaborated with US groups, while the radio astronomy research group at Manchester University developed ties with the optical astronomers in Manchester and elsewhere in the UK. The invitation of Fellows of the Royal Astronomical Society to Manchester in 1949 led to the formation of a committee on Radio Astronomy in 1950. The Royal Astronomical Society also began to support the plans for a big radio antenna at Jodrell Bank.

A main reason for the government support that was finally achieved was the independence of Radio Astronomy from weather conditions. Funding of £360,000 for this device was approved in 1951 by the Department of Scientific and Industrial Research. Other reasons for this decision included the high regard for British science due to its contribution to the war effort, the contribution of the project to national prestige (it was planned to be the largest radio telescope of the world) and the pioneering work of UK radio astronomers (both at Jodrell Bank and Cambridge) (Martin, 1978, pp 32–3; Munns, 2013, pp 118–9).

The recruitment of PhD students was a concern in all countries. In particular, the high prestige that physics (and especially nuclear physics) acquired during the war and the obscurity of Radio Astronomy in its first years made it very difficult for optical astronomy to attract students in the US, where it had to attract students from physics programmes. Radio Astronomy was much closer to physics and was successfully used by optical astronomers to attract students when the new radio telescopes began to operate at the Harvard and Caltech observatories (Munns, 2013, p 98). In England, where PhD students were funded by individual grants, both the Manchester and Cambridge groups were able to attract PhD students (Munns, 2013, pp 59, 86, 138). The Australian radio astronomy group was not part of a university and could only recruit graduated researchers (Munns, 2013, pp 84–5). The mobility of researchers between the Netherlands, UK, Australia and the US was already significant in the 1950s and contributed to the integration of Radio Astronomy and optical astronomy.

b) The experimental realization of Bose-Einstein Condensation as a process and complementary innovation

The nature of the innovation

Utilizing a major contribution to the new field of quantum statistics by Satyendra Nath Bose, Albert Einstein predicted in 1924 the existence of a new state of matter that occurs when gases of atoms are cooled to temperatures very close to absolute zero. This state, which is called Bose-Einstein Condensation (BEC), is characterized by particles losing their individual identities and coalescing into a single quantum state. Although widely accepted and explored as a theoretical phenomenon, the experimental realization of BEC was regarded by many physicists as being very difficult, if not impossible, to achieve for both theoretical and technological reasons. Theoretically, it was assumed for a long time that the interaction of particles 'would always lead to "ordinary" condensation (into a solid) before BEC would happen' (Ketterle, 2002, p 739). Technically, it was not clear how atom gases could be cooled down to the necessary low temperatures in a vacuum.

One of the first physicists to achieve BEC quotes reviewer reports on an NSF grant proposal that was approved in April 1995, that is, a few months before BEC was achieved:

> It is interesting to look at some of the reviewers comments now, seven years later: 'It seems that vast improvements are required [in order to reach BEC] … the current techniques are so far from striking range for BEC that it is not yet possible to make … an assessment …'; 'The scientific payoffs, other than the importance of producing a BEC itself, are unclear.' And a third reviewer: '… there have been few specific (or

> realistic) proposals of interesting experiments that could be done with a condensate.' Despite the skepticism, all reviewers concluded that the proposed 'experiments are valuable and worth pursuing'. (Ketterle, 2002, p 743)

In 1995, the first BEC of atom gases were experimentally produced by two US groups, which were later rewarded with the Nobel Prize. The methodological innovation at the time consisted in the development and combination of several cooling and imaging methods under vacuum conditions.

> Such an attempt needed a few years of very focused work as it involved the integration of several technologies that were not standard in the field, including ultrahigh vacuum, sensitive CCD cameras and image processing, high-current power supplies for magnetic traps, and flexible computer control of a multistep cooling and detection process. (Ketterle, 2002, p 745)

After experimental BEC was achieved, the research opportunities emerging from achieving experimental BEC beyond the application of the new cooling methods to other problems of low temperature physics were not immediately clear. Only with other groups first replicating these results and later achieving BEC in other atom gases it became apparent that BEC could be used to answer new questions of fundamental physics, and BEC was developed as a new field of research in physics.

Developing the innovation

The scientific community was initially undecided whether BEC would be the end of a long quest or whether it would open up new research opportunities. After experimental BEC was achieved by two groups in the US in 1995, it took other groups two years to replicate the results. The first experimental realization of BEC outside the US was achieved in Germany in 1997. This success was followed by many groups achieving BEC from 1998 onwards. Research has continued with attempts to create BECs of different atoms, as well as photons, solitons and so on. This still provides significant experimental difficulties, such as not all atoms being susceptible to laser cooling. For example, BEC in Neon has not yet been achieved after more than 20 years because Neon appears not to be susceptible to laser cooling.

The cooling methods combined for the manufacture of BECs were further developed and used in a wide range of experiments in low temperature physics. In addition, BEC itself became an experimental approach that has been used to study a wide range of problems in physics. With the manufacture of BECs becoming easier, more widespread and more diverse

(BEC in different atoms and later molecules being achieved), it became apparent that BEC can be used for a wide range of research problems in several subfields of physics, especially fundamental questions of quantum mechanics such as quantum many-body phenomena. 'A completely new field of research has emerged, in which ultracold atoms are used as elementary components in the investigation of fundamental properties of quantum matter' (Fallani and Kastberg, 2015, p 4). 'BECs have also been used to create atom lasers, atomic clocks and gravitational, rotational or magnetic sensors with excellent sensitivity' (Georgescu, 2020, p 396). Overall, BECs turned out to be very useful experimental platforms and BEC research grew rapidly to become an established field of research (Fallani and Kastberg, 2015).

Conditions supporting the development of the innovation

For a long time following the first experimental success, manufacturing a BEC remained a lengthy and risky process.[1] Experimental success even in the long time frames originally required strongly depended on research building on relevant traditions in the experimental physics fields on which the production of BECs was based, namely atomic and molecular optics, low-temperature physics and laser physics. Since BEC could be expected to occur only if a relatively dense atom gas was cooled to the lowest temperature possible at that time, researchers had to combine several recently developed advanced cooling methods. Among the advanced cooling methods required for achieving BEC, laser cooling was available only in few laboratories, and was 'spread' by postdoctoral researchers learning the method.

In addition to the theoretical and methodological knowledge, substantial material and personnel resources were necessary to achieve BEC for more than a decade after the first success. The necessary resources included a low-temperature laboratory as basic infrastructure and between 100,000 to 500,000 euros for additional equipment depending on the methodological difference between prior research and BEC. At least two researchers (postdocs or PhD students) were necessary to build and test the experimental set-up. Without access to substantial human and material resources, the experiments simply could not be conducted.

Achieving BEC in this period also took an unusually long time for an experiment in low-temperature physics. Constructing the necessary experimental system and understanding how it worked took several years. Most researchers who successfully manufactured BECs did so after more than five years, some of them needed eight years and more. During this time scientists were unable to generate significant publishable experimental results and so ran considerable risks of being regarded as unproductive. In some cases, journals did publish details of their experiments, but the reputational dangers involved in pursuing such long-term research projects that many physicists thought would result in failure remained considerable.

Faced with these risks, not all researchers interested in developing the BEC innovation did so (Laudel et al, 2014b). Several German researchers refrained because they feared not being able to compete with researchers from the US and especially with the two groups that announced their plans in 1993, succeeded in 1995, and whose results would not be replicated until 1997. Other researchers were less concerned about the competition but felt unable to marshal the necessary resources for the protracted period of time. These included two Dutch researchers on tenured positions below the professorial level who wanted to start BEC research but did not have access to the infrastructure. The professors who controlled this infrastructure were not interested in BEC research and did not allow the use of 'their' laser equipment for it. A newly appointed professor at a German university also could not access the necessary infrastructure because the university delayed the promised investment for several years.

Researchers who did decide to focus on BEC research depended on four conditions, namely: (a) significant material support by their university, (b) successive project grants, (c) the suspension of evaluations of their success and (d) approval of their plans by their peers acting as reviewers regardless of the majority opinion of their national scientific communities, which often were reluctant to accept the future potential of BEC. These conditions were met to different degrees and in different ways in five countries in which Laudel et al studied the development of BEC: Germany, the Netherlands, Spain, Sweden and Switzerland (Laudel et al, 2014b).

Material support by universities mainly depended on the career stage of the researchers attempting to achieve BEC. Established professors in all countries had their own laboratories, which some of them could convert for BEC research with loyalty packages.[2] Newly appointed professors received start-up packages with which they could equip a laboratory. A Swedish professor built his infrastructure entirely from a large grant for equipment (a kind of grant that did not exist in the other countries). In contrast, researchers below the professorial level needed to negotiate access to a laboratory. Several Dutch and German researchers were granted access to their professors' laboratories for BEC work. In addition to the equipment, university infrastructure for professors included discretion over positions for PhD students. Only the Swedish professor who built his infrastructure with a grant did not have discretion over personnel as part of the infrastructure funding.

Recurrent funding at universities was usually not sufficient to conduct BEC research. Only researchers at a German state research institute and two newly appointed professors in Switzerland were able to build the necessary protected space for their BEC research exclusively with recurrent funding.

Since the latter two had already conducted BEC research in their previous positions abroad, they were able to rebuild their experimental systems with generous start-up packages from their new universities and continue their BEC research lines.

All the other researchers needed to obtain project grants.[3] However, even during the early years following the 1995 experimental success, these were not easily obtained because the scientific community was divided about the utility of further manufacture of BEC. Researchers could, though, partially compensate for the reluctance of funding agencies and slow decision-making processes by bootlegging funds from other external grants. Funding agencies typically did not closely monitor the use of approved grants, which gave many researchers considerable leeway if they wanted to take up new lines of research quickly. Only the Swedish professor faced serious limitations because although his grant was approved for BEC research, it could only be used to fund equipment, and he was allocated only one PhD position by his department.

All researchers had to cope with the mismatch between the time needed to develop their experimental systems and the three or four-year terms of project grants and sometimes evaluations. Their peers who took up roles as assessors of grants or research performance at universities had to accept the lack of success after the common project cycles of three years. There was a remarkable difference in the attitudes of Dutch and German physics communities (and, consequently, grant funding agencies) towards these uncertain time horizons in the late 1990s and early 2000s. While German researchers received consecutive grants for BEC research regardless of their limited success with each single grant, the decision-making process in the Netherlands was different.

At the end of the 1990s Dutch researchers cooperated to submit a proposal for funding dedicated to BEC research. After reviews and revisions, which delayed the funding until 2000, this programme provided access to dedicated grants for researchers who wanted to produce BECs. However, when the programme was evaluated after three years, it was stopped because no further BECs were realized after the first success in 1999. The funding agency, FOM, then established a second funding programme for BEC but allocated all the funds of this programme to the group of the researcher who already had produced a BEC and to a junior research group leader in his department. These groups were the only two in the Netherlands whose research was not hindered by insufficient funding.

For some researchers, the long time required for successfully manufacturing BECs also came into conflict with evaluation cycles. For a Dutch researcher on a tenure-track position, this conflict was resolved by his faculty acknowledging the time frame of his research and suspending the mid-term evaluation. The Swedish researcher, however, found his department disappointed with his progress after the 'usual' time for projects to be completed.

Difficulties in gaining sufficient grants to deal with the uncertainties involved led two researchers to abandon their attempts to produce BECs. A Dutch researcher could not obtain grants for his BEC research anymore and had to give up. The Swedish researcher, who almost entirely depended on external grants, faced difficulties in securing sufficient resources to continue his research. In addition, he was criticized locally for his BEC research not producing enough publications, as the department did not allow for the long and unpredictable time horizon. As a consequence, he was not allocated any more PhD students from department funds and eventually he gave up BEC research and left Sweden.

The time and resource requirements have improved considerably over time. With the exceptions mentioned previously, 'making atomic BECs in the lab has become routine. There are commercial table-top BEC systems available and there are compact systems that can be used for portable sensors or free-fall experiments' (Georgescu, 2020, p 396). However, the further development of the innovation is still contingent on the general support by the state, universities and funding organizations for basic research without visible prospects for applications. If this support falters, as occurred in the Netherlands, the development of the BEC innovation might be constrained (Laudel and Weyer, 2014, pp 132–4).

c) Evolutionary-Developmental Biology as a complementary innovation

Nature of the innovation

Evolutionary-Developmental Biology (Evo-devo) is a field of research devoted to the question of how developmental processes and evolutionary change are related to each other. It emerged at the intersection of two fields: *evolutionary biology* (palaeontology, population biology) and *developmental genetics* (embryology). The innovation was initially conceptual. It responded to the difficulty of neo-Darwinian theory of the late 1970s and 1980s to account for all the empirical findings of evolutionary biology, particularly the rapid changes in the forms of organisms evident from the fossil record and the origins of traits that did not constitute an adaptation of the environment (Laudel et al, 2014a, p 236). As one biologist put it: 'It became increasingly obvious that these explanatory deficits of neo-Darwinism were due to its treatment of development as a "black box" and the consequent absence of the generative rules that relate between genotype and phenotype' (Müller, 2007, pp 500–1).

Evo-devo research opened this black box by empirically focusing on the evolution of the developmental processes that linked genotypes (genes) to phenotypes (structures, Hall, 2012, p 184). It was made possible by the development of molecular-biological techniques that enabled the empirical study of the molecular mechanisms of development. These methods and the discovery of the *Hox* gene family – genes regulating embryonic development – at the beginning of the 1980s made possible the comparative study of the development of different organisms.

Developing the innovation

The development of Evo-devo research is characterized by its integrative nature and its diversity. Instead of branching out into previously unexplored empirical domains, Evo-devo is a research field that integrates the previously separated fields of Evolutionary and Developmental Biology. Researchers from both fields have developed the innovation by moving to Evo-devo research. The costs of this move were generally higher for evolutionary biologists because they had to learn molecular-biological techniques and embryological methods, which most developmental biologists already knew and just needed to adapt to the new research questions (Laudel et al, 2014a, p 245).

Epistemic problems addressed by Evo-devo first centred on the evolution of developmental processes and particularly on the fact that all animals share gene sequences (the Hox genes) that:

> are responsible for determining that animals have an anterior and a posterior, a dorsal and a ventral side, and specific regions (often as repeated segments) along the body axis – head at one end tail at the other, thorax in front of abdomen, wings on a specific pair of segments, and so forth. (Hall, 2012, p 187)

Researchers discovered master genes that regulate development, and which are conserved across the Animal Kingdom (Hall, 2012, p 187). Later:

> an appreciation during the 1980s and 1990s of the importance of ecological and species interactions led evo-devo to the position where the aims of evo-devo could be stated as understanding:
>
> - The origination and evolution of embryonic development;
> - The role of modifications of developmental processes in the production of novel features;
> - How the adaptive plasticity of development facilitates the origin and maintenance of complex life cycles with embryos, larvae, and adults; and
> - How developing organisms interact with their ecological environment to facilitate evolutionary change. (Hall, 2012, p 187)

Currently, the development of Evo-devo includes epigenetics (the study of the causal factors that control gene action during development) and integrated studies using molecular biology, molecular genetics, developmental biology, phylogenetics, palaeontology and molecular paleobiology (which are revealing information about how features of organisms change during evolution). Future developments may include the integration of ecological, behavioural and medical questions in Evo-devo research (Hall, 2012, pp 188–9).

The innovation is developed mainly through comparative studies of genes regulating developmental processes in species located on the same or different evolutionary branches. The first wave of innovation development identified a set of genes regulating development that is conserved across most metazoans (multicellular animals). A second wave aims to identify how genes and molecules were used differently through evolution to build the past and present morphological diversity. A third wave has introduced the experimental testing of predictions drawn from the first two waves (Baguna and Garcia-Fernandez, 2003, p 705).

Contributions to Evo-devo research have been highly diverse. They varied in the extent to which research practices needed to be changed and in the resulting costs of such changes. Costs were lowest for non-empirical contributions, which included conceptual extensions, theoretical research and bioinformatics research. These contributions required no additional resources and only limited learning of new conceptual frameworks or skills. Conceptual extensions occurred when researchers used Evo-devo as an additional framework in which they contextualized their findings. They did not need to change their experimental research or the target audience for their empirical findings but added an interpretation for an additional target audience. Many researchers who later became fully engaged with the development of the innovation began their move to Evo-devo research with a conceptual extension of their previous research. Theoretical Evo-devo research differed from conceptual extension in that it was purely conceptual. Contributions were fully devoted to Evo-devo research but focused on conceptual development. The third kind of non-empirical contribution was mathematical modelling, which either used biological data produced by others or was based on more abstract models.

Empirical contributions to the development of Evo-devo greatly varied in their costs and in the extent to which they could advance the innovation. To make a substantial contribution to Evo-devo required comparative research of at least two organisms. Costs and benefits of Evo-devo research depended on how the information about the organisms was produced, and which organisms were chosen, as summarized in Table 5.2.

The choice of organism strongly affected the investments that were necessary for conducting Evo-devo research. Experimental biologists usually work with established model organisms, that is, organisms about which

Table 5.2: Costs involved in different kinds of experimental Evo-devo research

		Organisms used	
		Established model organisms	*New Evo-devo specific organisms*
Comparison strategy	*Internal* (experiments with two organisms)	High costs and risk for evolutionary biologists, lower costs and risk for developmental biologists	Highest costs and risk
	External (experiments with one organism, data about second organism from literature)	Lower costs, lower risk	High costs and risk

much knowledge has been accumulated over recent decades, which are therefore well understood, and for which experimental methods are known to work. However, these model organisms are not necessarily best suited for answering Evo-devo questions. For example, the evolutionary branch on which they are located constrains the choice of organisms they can be compared to. This is why answering key Evo-devo questions often requires working with entirely different organisms, which may be difficult to breed and for which the required methods might not work.

Necessary investments for experiments with two organisms further depended on the species studied. Researchers could use only one species and compare transgenic animals constructed from this species. They could use similar species (for example, two species of fish), which reduced the necessary learning and costs of breeding the animals. The highest costs were incurred when organisms from dissimilar species were used. The diversity of Evo-devo research questions has led to a proliferation of non-model organisms and new Evo-devo-specific model organisms, for which experimental methods are often difficult to establish (Sommer, 2009; Laudel et al, 2014a, pp 246–7).

Conditions supporting the development of the innovation

Requirements for developing Evo-devo strongly depended on the kind of contributions researchers intended to make.[4] Conceptual extensions, theoretical contributions and bioinformatics research did not require investments of resources or time beyond what was already available to researchers for their work in evolutionary or developmental biology. In contrast, experimental Evo-devo research could require substantial investments in facilities, staff and time, and in the early phase of the innovation its results sometimes had problems meeting community standards.

An important background condition for the development of Evo-devo research was the acceptance by intermediaries of purely basic research without the prospect of applications. This did not appear to be a problem because researchers who moved to developing the innovation came from evolutionary and developmental biology, which were also basic research fields. However, newly establishing Evo-devo research at universities depended on their continued support for basic research. This support became precarious in the Netherlands, where the overall pressure towards more applied research at universities created unfavourable conditions for Evo-devo research (Laudel et al, 2014a, pp 256–7; Laudel and Weyer, 2014).

Experimental Evo-devo research that did not use established biological model organisms could incur considerable costs because it required specific laboratories and infrastructure. In particular, breeding facilities for new species could be very expensive. As a consequence, this research was established primarily by particularly affluent universities like those in Switzerland and state-funded research institutes in Germany. German universities did not have the means to make the necessary investments. In Sweden, the costs of establishing experimental Evo-devo were covered by a large grant from a funder that was interested in establishing Evo-devo in Sweden (Laudel et al, 2014a, pp 257–8).

The long and unpredictable time horizons of experimental Evo-devo research restricted it to researchers on tenured positions, who also needed to control the necessary infrastructure. Together with the high costs of some forms of experimental Evo-devo research, these conditions appear to have limited the number of positions at which advanced Evo-devo research could be conducted. However, the move to Evo-devo research could be begun with non-experimental research, and researchers could choose less costly and less risky versions of experimental Evo-devo according to their circumstances. Indeed, Laudel et al (2014a) observed that for many researchers, non-experimental Evo-devo research provided the primary entry path into the field.

Until the new field of Evo-devo research was fully developed, evolutionary biologists and developmental biologists constituted the main audiences for this research. This became problematic for some Evo-devo researchers because those acting as intermediaries in these fields applied their standards to the contributions from the new interdisciplinary field, which could not always meet them. Laudel et al (2014a) report that some researchers had problems publishing in high-impact journals because they could not meet the standards of these journals. While dedicated Evo-devo journals already existed, their reputation and journal impact factors were lower. This created problems for Dutch researchers who had difficulties acquiring grants, and who could not meet expectations of their university concerning publications in high-impact journals (see also Laudel and Weyer, 2014).

Meanwhile, Evo-devo has become a strong and still growing research field which is supported by the intermediaries and audiences supporting basic research. While there is still research in Evolutionary Biology and Developmental Biology that does not belong to Evo-devo (Liberali and Schier, 2024), the partial synthesis of the two fields continues to be rapidly developed as an innovation.

Summary

These three case studies illustrate that although the development of innovations in practitioner-dominated science worlds depends to a large extent on the response by the scientific community, intermediary support was still important. The main difference to the cases discussed previously is that intermediary interests were one step removed from the content of the innovation and focused on some of its properties.

The rapid development of Radio Astronomy was supported by intermediaries which funded expensive new equipment – radio telescopes – under conditions of considerable scarcity after the Second World War. Their motives included the promotion of science as part of re-development of nations after the war, the intention to use radio telescopes as symbols of national scientific prowess and the wish to overcome limitations of traditional Optical Astronomy. They were thus directed less at the support of the innovation in question and more at the general support of a scientific field, with additional benefits in mind. Owing to these investments, Radio Astronomy was soon able to achieve the precision of measurements that convinced its main audience of optical astronomers of the utility of the new methods.

The cases of BEC and Evo-devo similarly highlight the role of practitioner communities and of generalized support by intermediaries. The initial reluctance of some national physics communities to support BEC was due to the unclear potential of the innovation. This reluctance soon faded once the potential of the innovation was clear. Evo-devo communities had no doubts about the utility of Evo-devo approaches and supported the innovation from the beginning, with some reluctance to lower scientific standards to accommodate new methods. The limited support for BEC and Evo-devo in some countries illustrates a background condition for the successful development of innovations in the fundamental sciences that is often overlooked, namely the willingness of state intermediaries to support expensive, long-term and risky fundamental research that does not promise applications.

Examples of stunted development of novelties

The cases discussed in the previous sections have in common that they represent innovations that were considered successful by those who developed them, their practitioner communities, and at least some intermediaries and audiences. The number of practitioners developing these innovations on an ongoing basis grew until each one became accepted and its development formed part of a practitioner community's standard repertoire. In order to emphasize the conditional nature of this kind of development, we now briefly consider two cases of stunted development – one in the sciences and one in the arts. In both cases, novelties initially attracted attention by some practitioners but subsequently failed to attract continued attention from the practitioner community. These developments remained marginal and diminished.

A prominent case of stunted development that was intensely discussed in the scientific communities involved and drew considerable attention from science studies is cold fusion (Pinch, 1994; Lewenstein, 1995; Simon, 1999, 2002). On 23 March 1989 Martin Fleischmann and Stanley Pons gave a press conference at which they reported the observation of excess heat (heat that could not be explained by chemical processes), and small quantities of above-background neutrons, gamma rays and tritium in an electrochemical experiment. They drew the conclusion that these observations indicated '[a] hitherto unknown nuclear process or processes' (Simon, 1999, p 65). It had to be an unknown nuclear process because no known process could explain the combination of measurements. This process was soon considered as an instance of 'cold fusion' (CF).

The announcement by Fleischmann and Pons was initially met with strong scepticism by the scientific community for two reasons. First, CF was considered impossible by the theory of nuclear processes. Nuclei would have to penetrate the Coulomb barrier (the repulsive force between similarly charged particles), which either requires the high energies at which 'hot fusion' is currently observed or quantum-mechanical effects, which occur at a much lower rate than the results by Fleischmann and Pons suggest. Second, if nuclear fusion occurred, much higher rates of fusion products (neutrons, gamma rays and tritium) should have been observed (Simon, 1999, p 66; Berlinguette et al, 2019, p 46).

Attempts to replicate the experiments failed in most but not all cases. A physicist who was invited to bring his own neutron detectors into Fleischmann's and Pons' laboratory reported detecting no above-background levels of neutrons (Simon, 1999, p 67). This report and failures of well-known and reputable scientists to replicate the experiments reported by Fleischmann and Pons quickly led the majority of the electrochemistry and nuclear physics communities to reject the findings. However, a closure of the controversy was not achieved. A workshop held by the US Department of Energy in cooperation with Los Alamos National Laboratory in May 1989 failed to produce a consensus on CF because some of the presented papers reported positive results (indicating CF) and others reported negative results (Simon, 1999, p 68).

Beginning with the second half of 1989, an increasing number of researchers publicly declared that they were abandoning CF research. In the US, public funding for CF research was shut down, beginning with the Department of Energy in mid-July 1989 (Simon, 1999, p 71). The majority of the scientific community began to consider CF as a case of pathological research (Simon, 1999, p 72). Nevertheless, CF research continued in some countries and received funding from public and private sources. For example, the Japanese government initiated the New Hydrogen Energy Research Project – a US$30 million programme directed towards the commercial development of CF technology – in 1993 (Simon, 1999, p 62). The main reason for this continued support by some intermediaries was the promise of CF research for providing cheap energy from abundant resources (Freire and Andrade, 2021, p 2). Conferences on CF continued to attract between 100 and 400 researchers for a long time. While the number of research groups engaged in CF research and the diversity of CF experiments continues to decline, there were still about 20 research groups worldwide that presented CF research results in 2020 (Freire and Andrade, 2021, pp 6–7).

Until today, continuing CF research remains characterized by two problems: its partial reproducibility and the lack of a convincing explanation. Until 2019, attempts to provide convincing evidence for the existence of CF failed (Berlinguette et al, 2019). Experiments reporting positive results for aspects of CF can sometimes be reproduced, while other attempts at reproduction fail. So far, little development beyond attempts to confirm the original claims by Fleischmann and Pons can be observed. The failure to replicate the originating novelty has prevented researchers from further developing it.

A similar case of stunted development appears to have occurred with aleatory, or random, approaches to constructing novels in the case of postwar experimental fiction in English. These approaches appeared less likely to influence the development of the novel than less radically divergent metafictional ones. As Patricia Waugh has suggested:

> The practitioners of so-called 'aleatory art' (which attempts to be totally random in order to suggest the chaotic, frenetic and colliding surfaces of contemporary technological society) are open to these charges (of ephemerality). Literary texts tend to function by preserving a balance between the unfamiliar (the innovatory) and the familiar (the conventional or traditional). Both are necessary because some

> degree of redundancy is essential for any message to be committed to memory. Redundancy is provided for in literary texts through the presence of familiar conventions. Experimental fiction of the aleatory variety eschews such redundancy by simply ignoring the conventions of literary tradition. Such texts set out to resist the normal processes of reading, memory and understanding, but without redundancy, texts are read and forgotten. They cannot unite to form a 'literary movement' because they exist only at the moment of reading. (Waugh, 1984, p 12)

Owing to the low number of studies of such cases of stunted development and to the lack of comparative investigations, it is impossible to identify the relative contribution of properties of the initial novelties, productive relationships between practitioners and authority relationships of art worlds and science worlds to these particular dynamics. A common property of both cases appears to be that the initial novelties did not offer sufficient opportunities for practitioners to develop them further. However, the case of cold fusion also suggests that such a limitation can be overcome if a development that does not meet the standards of relevance and/or quality of the practitioner community does appeal to the strong interests of intermediaries and audiences.

6

Developing Innovations in Changing Art Worlds and Science Worlds

These nine summaries of innovation development in differently organized art and science worlds have shown how the protected space needed for practitioners to undertake the development of, and gain acceptance for, substantial innovations in the modern arts and sciences was provided by a variety of combinations of practitioner actions, shifts in intermediary roles, policies and purposes, and changes in the nature and interests of audiences. These combinations often involved both unusual and infrequent changes in art and science worlds, such as the American Federation of Musicians' (AFM) national ban on members making records in 1942 and the later establishment of new firms that made and distributed recordings of Bebop music in the mid- and late 1940s, together with more frequent shifts in practitioner-intermediary roles, as when leading oncologists initiated significant changes in the National Cancer Institutes' priorities and funding decisions that supported the Oncogene Approach.

In this chapter we consider how changes in authority relationships governing aesthetic/epistemic priorities and standards as well as in the allocation of resources in art and science worlds have supported the acceptance and development of different innovations. In the first part we discuss how the nature of the particular combinations of changes in intermediary-led art worlds, practitioner-led art and science worlds, and practitioner-dominated science worlds provided the protected space for practitioners to develop and gain acceptance for different kinds of innovations and consider their broader implications for understanding how changes in the modern arts and sciences take place.

In the second part we consider how two sets of changes in the socio-economic conditions governing authority structures in many art and science worlds can be expected to affect the development of innovations in the arts and sciences. First, the increasing commercialization of creative production and distribution in art worlds and science worlds. The expansion of the international art market and associated intermediary roles and agents across the world extended investment-led art worlds and led to an emerging bifurcation of aesthetic goals and standards between those governed by a logic of artistic prestige and charisma and those dominated by a commercial logic. The commercialization of the sciences is leading to a rising number of science worlds with influential audiences and intermediaries and the inflationary growth of untrustworthy publications. Second, the expansion of state policies and agencies evaluating and guiding the use of public funds and associated forms of support that have shifted authority relationships in many art and science worlds.

The third part of the chapter summarizes the main points being developed in this book and draws some conclusions about the changing nature of novelty-driven fields of cultural production in the modern arts and sciences.

Changing authority relationships in art and science worlds

Comparing the nine case studies of innovation development and acceptance in the modern arts and sciences summarized in Chapter 5 enables us to identify the major shifts in different art and science worlds that provided the protected space for practitioners to undertake such commitments. In Table 6.1 we highlight the ways in which practitioner groups, intermediary roles and priorities and audience characteristics and interests changed the authority relationships governing novelty production, distribution and assessment such that practitioners' protected space was significantly enhanced.

a) Temporary weakening of intermediaries' dominance: Bebop and the Nouvelle Vague

In the two cases of innovations in the intermediary-led art worlds, Bebop and the *Nouvelle Vague*, there were a number of changes in the nature and priorities of intermediaries and audiences that together significantly reduced the authority of dominant intermediaries over both the selection of practitioners and the kinds of music and films that they produced and the prioritization of appealing to very large audiences. In the case of Bebop these included: the formation of the Broadcast Music Incorporated performance rights organization that later enabled black jazz musicians to become established as arrangers and composers of new music, the 1940s AFM ban on members making records that weakened the oligopolistic control by the three dominant companies and enabled new firms to be established that were more open to recording new jazz styles developed by

Table 6.1: Summary of nine innovations in the modern arts and sciences

Innovation	Type of innovation	Type of art or science world	Time frame	Contribution of changes in major authoritative agencies		
				Practitioner communities	**Intermediaries**	**Audiences**
Bebop	Displacement of jazz as dance music with jazz as concert music	Polycentric Intermediary-led	1940–50	Concentration of Black players in Harlem and emphasis on individual improvisation in small groups which could not be easily replicated	Club owners in Harlem providing opportunities to experiment and access to small audiences. New record firms, radio stations, supportive disc jockeys and agents link innovators to new national audience	Jazz as music to listen to (rather than dance music) wins new urban and later national audience
Nouvelle Vague	Complementary	Polycentric Intermediary-led	1958–70	Critic-auteurs developed skills making short films and documentaries, made features often with lower cost practices and improvisation	Postwar state becomes authoritative intermediary for French cinema. Protected space through state funding for experimental film-making	New youthful audience for films addressing their concerns, rapid decline of mass heterogenous audiences in late 1950s
Cubism	Displacement of traditional techniques with a new pictorial language and aesthetic	Polycentric practitioner-led	1907–20s	Salon Cubists provided exhibition and communication opportunities in enlarged avant-garde practitioner community in Paris	Unconditional support for Braque and Picasso through dealers and some investors/collectors. Supported by literary avant-garde and Salon Cubists taking on intermediary roles	Extension of avant-garde audiences through international exhibitions and of collectors for new artistic approaches

(continued)

Table 6.1: Summary of nine innovations in the modern arts and sciences (continued)

Innovation	**Type of innovation**	**Type of art or science world**	**Time frame**	**Contribution of changes in major authoritative agencies**		
				Practitioner communities	**Intermediaries**	**Audiences**
Metafiction	Complementary	Polycentric practitioner-led	1960–70s	Academic writers at universities become writer–critics and gain authority in academically dominated literary field	Publishers support novel narrative forms and university positions for language-focused authors grow. Shift of authority to university-embedded social-professional networks connecting authors to editors and publishers	Expansion of novel reading public and college student audiences for deviant styles, boosted by growing creative writing programmes and counter-cultures
Oncogene Approach	Displacement of object-centred shared background with Oncogene Approach and molecular-biological methods	Polycentric practitioner-led	1970s and 1980s	Rapid acceptance of Oncogene Approach due to supporting empirical evidence and of molecular-biological techniques due to new research opportunities provided	Intermediaries provide immediate financial support Increased private investment in new methods	Support by cancer researchers who benefited from oncogene theory and funding of molecular-biological methods

Table 6.1: Summary of nine innovations in the modern arts and sciences (continued)

Innovation	Type of innovation	Type of art or science world	Time frame	Contribution of changes in major authoritative agencies		
				Practitioner communities	**Intermediaries**	**Audiences**
Financial Economics	Displacement of firm-centred and institutional problems by market- centred conceptual approach of neoclassical economics	Polycentric practitioner-led	1960s and 1970s	Epistemic authority of neo-classical economics legitimated innovation New research opportunities due to changes in corporate finance and digitalization of financial operations	Business school support for 'scientific' teaching and research. Leading journals, foundations, patrons and universities supported approach	Capital market actors and leading economists supported approach. Nobel prizes for leading financial economists
Radio Astronomy	Process	Integrated practitioner-dominated	1940s and 1950s	Communication and collaboration of radar physicists on new observations, early collaborations with some optical astronomers	Academic employment of radar physicists using leftover radar equipment from WWII. State agencies promoting basic research, patrons and foundations all supported Radio Astronomy with major investments in equipment	Development of techniques and results gained support of optical astronomers and were integrated into general astronomical research
Bose-Einstein Condensation in atomic and molecular Optics	Process to 2010s Complementary from 2010s	Integrated practitioner-dominated	Since 1995	After initial reluctance widespread support when experimental practices were standardized and epistemic value was demonstrated	Support by intermediaries despite scepticism concerning possibility of success and epistemic value. Employers and funding agencies in many countries support long-term, risky and expensive basic research	After some delay due to reproduction difficulties, strong support from wider physics community following new research opportunities

(continued)

Table 6.1: Summary of nine innovations in the modern arts and sciences (continued)

Innovation	Type of innovation	Type of art or science world	Time frame	Contribution of changes in major authoritative agencies		
				Practitioner communities	Intermediaries	Audiences
Evolutionary-Developmental Biology	Complementary	Integrated practitioner-dominated	Since early 1980s	Practitioners attracted by opportunities to solve growing problems of neo-Darwinism with rapidly diffusing molecular-biological methods. Opportunities for 'low-cost' approaches	Employers and funding agencies in some countries supported long-term, risky and expensive basic research, some initial reluctance of journals to accept contributions not meeting established standards	Strong support by both evolutionary and developmental biologists, although both fields retain their independent identities

small combos of Black musicians, the postwar expansion of radio stations and jukeboxes together with novel intermediary roles such as disc jockeys that helped to establish national audiences for listening to modern jazz in concert halls, festivals and other novel locations rather than dancing to swing styles.

In the case of the French *Nouvelle Vague*, the longstanding elite support for the cinema as an artform, rapid postwar expansion of *ciné-clubs* and specialist critical journals coupled with continued state support for the domestic film industry in competing with US imports limited the dominant role of oligopolistic production and distribution firms in determining the nature and purposes of filmmaking. Particularly after modifying the broad government policy of supporting film production and distribution in France as a whole towards funding more innovative projects in the mid-1950s, the state significantly reduced the risks involved in investing in unconventional projects focusing on the concerns of particular audiences. Insofar as these risks involved relying on young, relatively unknown actors, cinematographers and directors, they were also mitigated by the success of both foreign and domestic films appealing to more youthful audiences than the mass audiences previously sought by intermediaries.

In both cases, the initial stages of innovation development were supported by the protected space in which practitioners were able to develop their technical skills, aesthetic goals and cultural identity in relative separation from the dominant community. By taking advantage of the autonomy granted to the house band and proficient visitors at after-hours clubs in Harlem, Bebop musicians were able to develop a distinct set of skills and practices that were beyond the competences of most outsiders, particularly white musicians who otherwise would probably have dominated the commercialization of the novel style themselves. Similarly, postwar support for the production and exhibition of short films and documentaries in the late 1940s and 1950s provided opportunities for aspiring directors to gain experience in the French film industry, and ready access to both magazines focused on the cinema and more generalist media enabled them to develop and publicize their distinctive aesthetic purposes and identities.

Taken together, these two cases suggest that in intermediary-led art worlds, innovations which change the aesthetic problems of practitioner communities by replacing some of them or by adding new ones can be successful if the authority of incumbent intermediaries is temporarily weakened by competition from new intermediaries who create or grant protected space for practitioners developing the innovations. Incumbent intermediaries had little interest in development of innovations whose commercial potential was not clear. The weakening of their authority by new intermediaries was supported by the emergence of new audiences to whom they had limited access.

b) Alignment of intermediary priorities and audiences with practitioner interests: Cubism, Metafiction, Oncogene Approach and Financial Economics

In the four cases of innovations in practitioner-led art and science worlds, Cubism, Metafiction, the Oncogene Approach and Financial Economics, there were a number of shifts in the nature and priorities of intermediaries and audiences that had significant effects on the protected space available for practitioners developing substantial innovations. In all four cases, intermediary agents directly or indirectly provided financial support for their work and pursued common interests with them. For Cubism these included the willingness of some art dealers to support painters pursuing highly unconventional approaches to longstanding aesthetic problems by committing themselves to buying all their new works over a given period as well as sometimes agreeing to provide living expenses. Some, such as Kahnweiler, also prioritized the exhibition of new works abroad rather than submitting them to the national Salons in order to generate an international reputation to counter local conservatism. This internationalization of exhibitions of avant-garde works helped to develop the pan-European avant-garde network of artists in the decade before the First World War.

Additionally, although most journalist commentary on avant-garde movements in Paris from the Impressionist exhibitions in the 1870s and 1880s onwards was negative, if not derisory, the expanding number of little magazines that commented on experimental works in the arts provided opportunities for the literary avant-garde to support new approaches in painting. Together with the expansion of the private *academies libres* and dealer-centred art market, they helped to sustain an artistic avant-garde social formation that was dedicated to aesthetic disruptions by the end of the 1890s. This willingness to take risks in supporting the production of unconventional works by young artists after the turn of the century was additionally supported by the expansion of the audience of investors acting as the 'winklers out' of new works by less well known artists, as exemplified by the Bearskin syndicate, which followed the eventual market acceptance of the Impressionist movement and somewhat more rapid recognition accorded to the post- or neo-impressionists.

Comparable changes in the conditions supporting the production, distribution and assessment of Anglophone fiction occurred in the 1960s and 1970s. Following the expansion of the market for novels, partly generated by the paperback revolution, publishers became more willing to support experimental narrative forms. Additionally, the postwar expansion of tertiary education and extension of state support of academic research to the humanities enlarged graduate education programmes in departments of English as well as supporting the academic integration of practitioner and critic roles. Together with the declining authority of professional critics and elite cultural networks, this enhanced the ability of novelists to pursue unconventional approaches to the orthodox presentation of narratives and helped to develop novel forms of literary criticism. Intermediary authority partially shifted from established intermediaries to the new author-critics in the increasingly academically dominated literary field.

The parallel growth of creative writing programmes in many English departments provided additional support for novelists developing metafictional approaches, although some of these programmes had previously followed more traditional analytical frameworks. As well as offering employment opportunities for writers, they enlarged the size of the knowledgeable audience for new fictional styles, which in turn helped to encourage publishers to invest in them. More generally, the expansion of the college student population and development of counter-culture movements during the Vietnam war broadened the audience for unconventional novels, particularly when they were humorous and overtly critical of dominant approaches and assumptions as well as of established elites.

In the case of the Oncogene Approach, the limited success in finding viral causes of cancer following the concentration of research funding on viral agents in the National Cancer Plan of 1971 and the considerable practitioner criticism of this exclusive focus put pressure on intermediaries. The discovery of oncogenes and proto-oncogenes enhanced the authority of elite researchers in reorienting science policy priorities. As a result, the administrators of the National Cancer Institute broke up the Viral Cancer Program and refocused funding on the use of molecular-biological methods to analyse oncogene development and develop ways of intervening in this process. Together with considerable private investment in providing resources for the development and standardization of molecular-biological methods, tests and software for oncogene research, this reorientation of epistemic priorities and research practices was quite rapidly accepted by the primary audience of practitioners researching the wide variety of exogenous causes of cancer including spontaneous mutations, chemical compounds, hormones, and radiation or viruses, as well as by the larger audience of politicians, the health care industry and the public.

The transformation of business finance research and teaching into Financial Economics in leading North American business schools between the 1960s and 1980s depended on similar major changes in the nature, priorities and authority of intermediaries and audiences. Particularly important was the increasing competition between university business schools for academic recognition of the research produced by their faculty following growing pressure from philanthropic foundations, business groups and some employers to contribute to the development of the management sciences in dealing with the increasingly complex problems faced by large, often multidivisional companies. As well as the rapid expansion of MBA programmes, this led to the growth of doctoral programmes whose qualifications became required for business school faculty. This expansion of research focused staff encouraged the creation of new academic journals specializing in new subject areas and approaches and the establishment of new research communities competing for recognition and resources in the enlarged business schools.

This competition depended on the interests and status of two sets of audiences: established academic disciplines and business elites, including employers. In the 1950s and 1960s, mathematics and the physical sciences were particularly highly regarded academic fields in most Organisation for Economic Co-operation and Development (OECD) countries and the new management sciences were often thought to contribute most by developing mathematical models that could provide optimal solutions for business problems. However, the social sciences were more directly relevant academic audiences for most business school faculty seeking intellectual reputations in research-based universities and economics was the most authoritative of these fields, as indicated by the Nobel Memorial Prize in Economic Sciences, dominated by neo-classical presumptions of efficient markets and fully rational actors. Focusing on inter-period price movements in capital markets best understood as a random walk model, with the established intellectual apparatus of orthodox economics was a natural preference for those developing Financial Economics.

The creation and analysis of data bases of stock market price movements was supported by some investment banks, such as Merrill Lynch, and other financial businesses, together with some business owners who believed in the natural efficiency of capital markets and wished to reduce government regulation. Additionally, the growth of collective investment management organizations, such as mutual funds, unit trusts and pension funds, in the 1960s and 1970s generated concerns about their management of financial risk, increasing the audience for modern portfolio theory and the development of techniques for measuring risk.

In all four cases, incumbent resource-providing intermediaries actively supported the innovation by investing resources in protected space for the practitioners developing it. Developing the innovations was in the interest of these intermediaries because they were likely to be successful with extending audiences, be it a growing international audience for Cubism, the growing number of university students at creative writing programmes as an audience for metafictional approaches, the extended number of cancer researchers that could develop novelties based on the Oncogene Approach or the priorities of economists and investors. Further support for Cubism, Metafiction and the Oncogene Approach was provided by practitioners who additionally took on roles of interpretive intermediaries. These role hybridizations additionally aligned intermediaries' interests with those of practitioners developing the innovation.

c) Intermediary support for practitioner-dominated sciences: Radio Astronomy, Bose-Einstein Condensation (BEC) and Evolutionary-Developmental Biology (Evo-devo)

In the three cases of innovations in Integrated practitioner-dominated science worlds, Radio Astronomy, BEC and Evo-devo, the successful development of innovations depended on support by practitioner communities and intermediaries for what was basic research that took a long time, was risky and expensive. In the case of Radio Astronomy, early contributors were usually employed by universities and often able to recruit PhD students to work on radar equipment left over from the Second World War. Later on, in the UK Bernard Lovell went to considerable lengths to gain the support of senior optical astronomers when applying to the government for support in building the world's largest radio telescope and, more generally, collaboration with established optical astronomers increased as radio astronomical techniques and results became more precise. The willingness of state agencies to support the innovation reflected the independence of radio astronomical observations from changing weather conditions as well as their contributions to the physical sciences and other government priorities.

In the case of BEC, funding for the difficult and expensive experimental research was initially supported in the USA by reviewers for the National Science Foundation (NSF) despite widespread scepticism about its feasibility in Atomic and Molecular Optics. Once successfully accomplished in 1995, support for similar projects was also provided in a number of European countries, but success often proved elusive, partly because it depended on longer term commitments by employers and senior professors as well as by state agencies and their reviewers and other intermediaries such as journal editors. This case highlights the importance of gaining continued support from multiple agencies, practitioner communities and organizations at different levels of collective action for developing long-term, expensive and risky innovations involving the elaboration and coordination of different skills and knowledge over longer periods than the usual funded project and evaluation cycle. After over many years of such support for BEC research, the innovation became the means for investigating a wider range of problems in physics.

The development of Evo-devo also illustrated the need for continued financial and other modes of collective support for the development of long term, expensive and risky innovations. The combination of dissatisfaction with neo-Darwinian theories that failed to account for all the empirical findings of evolutionary biology, the rapid diffusion of molecular-biological methods that enabled researchers to study the molecular mechanisms of developmental processes and the discovery of the Hox gene family that regulate embryonic development encouraged the comparative analysis of developmental processes in different organisms in largely conceptual forms of Evo-devo research.

However, comparisons of long-term developmental processes in non-model organisms dealing with major Evo-devo questions were often dependent on considerable investment in animal houses, laboratories and infrastructure specific to these organisms, particularly if they were from different evolutionary branches. Justification for such investments typically depended on continuing support from both developmental and evolutionary biologists advising national research councils and those controlling resources and assessments of the quality of results at employment organizations and leading journals. Overall, the more developing an innovation depended on long-term investment in learning a variety of new practices and knowledge of new phenomena and conditions, the more it rested on continued practitioner, intermediary and audience support at international, national and local organizational levels.

The three cases of innovations in practitioner-dominated science worlds demonstrate that the development of the innovations was under the authority of practitioners who simultaneously were the dominant audience appreciating the innovation. Nevertheless, the development of the innovation also depended on intermediary approval because scientific communities depend on funding by the state and thus on the general approval by its intermediaries of work on the epistemic problems shared by these communities. As the limited approval of the work on BEC and Evo-devo by Dutch intermediaries demonstrated, even in practitioner-dominated science worlds with their coinciding practitioner communities and dominant audiences developing major innovations is still contingent on general intermediary support.

The nine case studies suggest a number of connections between particular conditions and the development of particular innovations. With regard to the role of authority structures in practitioner communities, they demonstrate that displacement innovations that replace the shared background of Polycentric practitioner communities are possible (Bebop, Cubism, Oncogene Approach and Financial Economics), while complementary innovations with a narrower focus add new approaches to existing ones (*Nouvelle Vague*, Metafiction). Integrated practitioner communities in the sciences support process innovations and complementary innovations as soon as they agree on their epistemic utility (Radio Astronomy, BEC, Evo-devo). The same is likely for displacement innovations in Integrated practitioner communities, for which there was no case study.

The development of different kinds of innovations also seems to be connected to variations in authority relations in art worlds and science worlds. In intermediary-led art worlds, the development of a displacement and a complementary innovation was facilitated by a temporary weakening of incumbent intermediaries and the shift of authority to new intermediaries that linked practitioners to new audiences (Bebop, *Nouvelle Vague*). In practitioner-led art worlds and science worlds, the support of incumbent or new intermediaries for the particular complementary or displacement innovations was a major supporting factor, along with the extension of audiences including the members of scientific communities in their audience roles (Cubism, Metafiction, Oncogene Approach, Financial Economics). In practitioner-dominated science worlds the development of displacement, process and complementary innovations, while under the control of practitioners, still depended on supportive audiences and intermediaries even if the support was not directed at the specific innovation but at the practitioner community more widely (Radio Astronomy, BEC, Evo-devo).

Finally, an important contribution of the case studies is their portrayal of changes in innovations and practitioner communities themselves. In the case of BEC, what began as a process innovation turned into a complementary innovation as practitioners moved to a rapidly developing area of new research opportunities. The Oncogene Approach increased the integration of the field of cancer research and transformed it from a *Polycentric* to an *Integrated* practitioner community. The same happened to Financial Economics, albeit only temporarily. The displacement of a firm-centred by a market-centred approach integrated the field for more than a decade before divergent developments of approaches made it polycentric again. These examples demonstrate that the development of innovations may change their character and the structure of the practitioner communities which develop them.

Changes in the social conditions governing authority structures in art and science worlds

This discussion of conditions for developing different kinds of innovations in modern art worlds and science worlds considered these as stable social contexts for innovations. This abstraction was necessary to identify the influence of authority relations on the development of different kinds of innovations. However, the socio-economic conditions under which art and science worlds exist frequently change, as do the social structures and authority relations in art and science worlds. We now briefly discuss some examples of major changes currently affecting them and possible consequences of these changes for the distribution of authority in art worlds and science worlds. One major development in art worlds is the rapid expansion of transnational markets in contemporary art works, which has initiated an emerging bifurcation of aesthetic goals, standards and authority relations. For the sciences, major changes in their governance and funding have taken place with some consequences for changing authority relations in science worlds and opportunities to develop innovations.

a) Changing conditions in the contemporary art world of plastic arts

The conditions in which the production, exhibition, recognition and sale of contemporary art works are carried out have changed considerably since the 1980s. The market for contemporary art works has greatly expanded across the world as artists, dealers, auctioneers and collectors increasingly produce, sell and buy them in a variety of venues and organizational settings in different countries and continents, including commercially oriented art fairs, transnational contemporary art dealers and galleries and public auctions of contemporary art works. Similarly, the network of stakeholders involved in the development, exhibiting, viewing, discussing and evaluating of contemporary art has broadened to include new roles and institutions and redefine established ones. These include transnational contemporary art biennial exhibitions, other large-scale non-commercial contemporary art exhibitions drawing on contributions from many countries, new public and private museums of contemporary art, freelance contemporary art exhibition curators, new contemporary art critical reviews and means of ordering the aesthetic merits of different contributions (Acord, 2010; Horowitz, 2011, p 12).

Together with the increasing number of contemporary art collectors in rapidly growing economies, these intermediaries can be seen as contributing to the construction of a distinct global art world in which the global market in the production, display and exchange of contemporary art works is absorbing the previously dominant 'North-West' art world of artists, critics, galleries, curators and collectors based in North America and Europe. However, the continued divergence of artists' aesthetic reputations and market success in the 21st century together with the greater cultural and aesthetic diversity of artistic production, exhibition and recognition in the exhibition focused contemporary art world than in the commercially dominated art world, suggests that these developments represent more *an emerging bifurcation of art worlds* than an absorption of the previous one by commercial logics (Lee 2018; Buchholz, 2022; Brown, 2023).

Considering first the expansion of *Biennale* exhibitions of contemporary art, these increased dramatically in the 1990s and 2000s to reach 130 non-commercial and regularly repeated exhibitions that featured artists from at least three continents, typically North America, Europe and one or more other. By 2017, 40 per cent of these took place in North America and Europe (a significant drop from the 67 per cent in 1980), 32 per cent in

Asia, 11 per cent in Eastern Europe, 8 per cent in Latin America and 7 per cent in Africa, with curators also becoming much more diversified in terms of their national origins (Buchholz, 2022, p 32). While many of these new *Biennale* exhibitions were intended to put a host city:

> on the art world map and, at the same time, to increase the visibility and symbolic capital of the art of the country or region where they are located … A biennal could also serve as a cultural vehicle supporting postcolonial nation building, as an instrument of Cold War politics by means of an international exhibition, as an event for the marking of political transitions, and as a means of bolstering tourism and urban development to cities now promoted to cultural capitals. (Buchholz, 2022, p 33)

Their curators also became much more diversified in terms of their national origins and often presented novel aesthetic approaches in their selection and organization of works, almost acting as 'authors' of new understandings of artistic development and recognition (Buchholz, 2022, p 200).

More permanent exhibition institutions for contemporary art also proliferated across the globe in the 1990s and 2000s. The Guggenheim museum, for instance, opened branches in Bilbao and Berlin in 1997 as well as already managing the Peggy Guggenheim collection in Venice. According to Buchholz (2022, pp 286–8), by the mid-2010s there were almost 4,000 non-commercial art museums, public galleries, contemporary art centres, *Kunsthallen*, art associations and university galleries that included foreign-born artists in their contemporary art exhibitions.

In general, artists exhibiting in these institutions have adopted much broader conceptualizations of what constituted a contribution to the visual arts than the easel paintings and figurative sculptures that dominated previous art worlds and were more valued by artists pursuing commercial success (Buchholz, 2022, pp 139–42). These included video art, installations, performance art, body art and similar 'dematerializations' of the art object that followed the advent of conceptual art and postmodernism. These forms of experiential art emphasized the authorial role of artists and the extension of copyright legislation to the sale, presentation and conservation of art works. In this understanding, collectors, rather than acquiring unique, tangible goods, are buying:

> systems of rights, expectations , and obligations governing how these goods are to be presented and conserved … (when) Steven Cohen bought (Damian) Hirst's *Physical Impossibility of Death in the Mind of Someone Living* (1991) … he did not really purchase a static sculpture (the original shark-in-tank), but the intellectual property that gave birth

> to that sculpture as a legitimate Hirst artwork and which would allow Cohen to present it as such and to renew it as necessary. (Horowitz, 2011, p 105)

This extension of artists' rights and control over the disposition, exhibition and maintenance of their works represented a considerable enhancement of their autonomy and authority over intermediaries and audiences.

Turning next to the rapid expansion of the commercial part of the contemporary art world, the initial rise of the 'mega' art galleries, such as Gagosian, and their establishment of highly valued artist-gallery brands comparable to those in the fashion industry, mostly took place in the 1980s in the USA. However, the larger-scale expansion of the contemporary art market across the world occurred in the late 1990s and early 2000s and is often attributed to the post-Cold War rapid growth in the population of high net-worth individuals (HNWI) who had over US$1 million to invest. This number rose from 5 million in 1996 to over 15 million in 2016 and the number of billionaires in the world increased from around 500 in the mid-1990s to 2000 in the mid-2010s (Buchholz, 2022, p 76).

Almost in parallel with this growth of the wealthy, was the establishment and expansion of a novel form of exhibiting and selling works of contemporary art – the international contemporary art fair (ICAF) (Quemin, 2013). The first of these was established in Germany by local dealers in 1967, but they became much larger and with more participation by high profile galleries from North America and other European countries in the 1990s. The fastest rate of growth of ICAFs took place in the first decade of the new millennium, with the total number of transcontinental fairs reaching 126 by 2017 (Buchholz, 2022, pp 75, 289–90).

In 2017 the bulk of these (72 per cent) were held in the 'North-West' of North America and Western Europe, with a further 16 per cent taking place in Asia, particularly Hong Kong and China. While many of the fairs were either set up or taken over by the largest ICAF groups such as Art Basel, others were often the fruit of local initiatives in Latin America, Africa and elsewhere, whether private or governmental. As they expanded, their ownership and control tended to change from galleries to major trade fair companies primarily concerned with expanding operations to newly wealthy locations and attracting the largest and most well-known galleries and collectors (Buchholz, 2022, pp 73–84; Horowitz, 2011, pp 131–9). Increasingly, sales at ICAFs are becoming a significant proportion of galleries' total sales, accounting for one third of total dealer sales in 2013 (Gnyp, 2020, p 148).

This growth of ICAFs in recent decades both reflects and affects changes in the nature and role of art galleries across the world. Although the prominence of the radically sales-maximization orientation of the dealer–artist brands declined somewhat in the 1990s after the 1989 recession, it became more widespread and influential in the 2000s as the number and commercial success of ICAFs grew significantly. While this commercial success made the fairs more attractive to galleries and encouraged them to seek art collectors across the world, the costs of participating in them rose considerably and limited the ability of smaller galleries and those preferring to support aesthetically challenging artists to attend, leading to their domination by the better capitalized galleries. Their short duration and often restricted spaces also encouraged galleries to show works that were moderate in size and likely to fit in with current commercially attractive styles, mostly paintings and traditional sculptures rather than videos, installations and performance art (Buchholz, 2022, pp 73–83; Gnyp, 2020, pp 148–53).

Another change in the contemporary art world that has weakened the authority of aesthetically focused intermediaries and collectors is the increasingly influential role of the major auction houses, Christie's and Sotheby's, in establishing the market value of living artists' works. Before the 1990s they had tended to focus on Impressionist and Modern artworks rather than those by living artists. Between 1998 and 2008 however, the worldwide auction turnover of contemporary art works by artists born after 1945 rose from US$48 million to US$1,349 million in 2008 with Sotheby's sales of contemporary works made after 1945 reaching US$1.3 billion (including the buyers' premium) in 2007 and Christie's reaching US$1.6 billion (including the buyers' premium). In both cases, these contemporary art sales exceeded those in the Impressionist and Modern art category, partly because the supply of high-quality older artworks in circulation had diminished as more became absorbed by museums and private collections (Horowitz, 2011, pp 7–9, 193).

These two auction houses also played a major role in expanding the market for contemporary art as they pursued the growing population of wealthy collectors in Asia, the Middle East, Latin America and elsewhere in the 2000s and 2010s. Whereas most of their sales were held in London, New York and Continental Europe before the 1980s, and were dominated by Impressionist and Modern art works, Sotheby's held auctions in Hong Kong from 1973 onwards to be followed by Christie's in1984 and both operated representative offices in Australia, East Asia and Latin America as well in cities across Europe and North America by the end of the 1980s when they began to focus more on contemporary artworks. After 2000, they both started to hold auctions in Paris, Dubai, Doha, Shanghai and Mumbai (Buchholz, 2022, pp 84–94). As collectors rather than dealers have become the primary clients of the auction houses, they have established financial services units to advise collectors and offer them credit loans, price guarantees, private sales and similar services as well as publishing more detailed sales brochures and linking works for sale to established masterpieces to tacitly boost their appeal and auction value (Gnyp, 2020, pp 161–7).

Together with the expansion of private museums of contemporary art and the wealth of many collectors, such as Bernard Arnault and Francois Pinault, these developments can be seen as enabling private galleries, dealers, auction houses and collectors to exercise increasing aesthetic authority over the contemporary art market and the judgements of public museums, curators and critics (Quemin, 2020). However, it remains to be seen whether this expansion of commercial interests and authority over aesthetic standards would be sustainable in the event of a major collapse of both the primary market for new works and the secondary market for already bought works or, indeed, significant downturns in transnational economic growth.

b) Changing conditions in science worlds

In most countries, the sciences, social sciences and humanities have grown rapidly since the Second World War, mainly because scientific knowledge has become a major source of economic growth, with an increasing number of industries exploiting scientific results, and because the increasing dependence of societies on the sciences has driven the demand for higher education. The number of publications has grown exponentially at an annual rate of approximately 8–9 per cent since 1980 (Bornmann and Mutz, 2015; Fortunato et al, 2018). The number of researchers also grew exponentially, albeit at a slower annual rate of about 3–4 per cent (Johnson et al, 2018).

The increasing dependence of modern societies on the sciences and the increasing costs of maintaining growing science systems have led to significant changes in the governance of science at different speeds in different countries. Six changes are of particular importance because they affect the interests of audiences and intermediaries concerning science and the distribution of authority in science worlds (Whitley, 2010). First, research funding transitioned from a mode dominated by recurrent funding for research organizations to a split funding mode in which shrinking recurrent funding was increasingly complemented by competitive grant funding. In many fields, researchers must win grants to be able to conduct research. Public and private grant funding agencies have emerged as new intermediaries, with state-funded research councils becoming the most important source of money for research projects (Rip, 1994), albeit at different times in different countries. The state and scientific communities usually share authority over the distribution of grants, with the authority distribution between the two varying between countries (Braun, 1993). The growth of science has led to increasing competition for grant funding, with success rates for grant funding diminishing in many countries.

Second, public policy goals have become much more influential in setting research priorities. The growing dependence of societies on scientific knowledge has strengthened the state's interest in steering the directions of research, which state actors pursue by tying funding for research to public policy goals and by using other policy measures like performance agreements with universities to exercise influence on researchers (Guston, 2000; Drori et al, 2003; Hessels et al, 2011).

Third, public research organizations (PROs) in many OECD countries have developed greater strategic autonomy as organizations. While the rhetoric of higher education reforms tends to overestimate the degree of this newly won autonomy (Capano, 2011; Gläser and Schimank, 2014), the development of specific organizational interests as actors and the enhanced capability to act according to these interests has increased their authority as intermediaries in many European countries.

Fourth, the growing dependence of the economy and society more generally on the sciences has increased the opportunities for, and the encouragement by the state of, research commercialization by scientists and PROs. The state's interest in promoting the national economy, the economy's dependence on science for technological innovations and new opportunities for PROs to benefit financially from these innovations all contribute to a greater influence of both PROs and private business interests on research goals. These initiatives have stimulated a growth in university applications for patents based on academic research and generated novel modes of private investment in academic work (Kearnes and Wienroth, 2011; Berman, 2012).

Fifth, PROs have become more subject to systematic, formal and consequential state monitoring of their performance (Whitley and Gläser, 2007; Hicks, 2012). This monitoring is in part an instrument for implementing public policy goals, in part an attempt to demonstrate the efficiency of public funding and – in the case of competitive performance-based funding of PROs – an attempt to legitimize insufficient funding for science and higher education. It is thus in many cases not an exercise of epistemic authority but rather an exercise of other forms of authority that may have epistemic consequences (Gläser and Laudel, 2016).

Sixth, the increasing importance of scientific knowledge for national welfare and security has given rise to techno-nationalism, that is to policies of subordinating national knowledge production to political or economic interests. Similar to some governance measures for performance measurement, techno-nationalism does not imply the exercise of epistemic authority but has epistemic consequences. The most influential case in point is US export controls, which have affected international scientific communities since onset of the Cold War, particularly in the context of embargoes against the Eastern bloc until the late 1980s. Today, export controls affect both the movement of researchers through visa regulations (Daniels, 2019) and the movement of knowledge through collaboration and publication (Krige, 2019). Both pathways affect international scientific communities, for example through hindering communication and collaboration (Britz and Ponelis, 2012; Varma and Sabharwal, 2018; Wondimagegn et al, 2022; Lee and Li, 2023).

In addition to changes in the governance of science, authority structures of science worlds are also affected by changing practices of communicating scientific results. The growing pressure to publish can be traced back to the increasing competition for positions and resources and to the actions of public research organizations, which in response to the monitoring of their performance demand more publications from their researchers. As a consequence, a growing number of masters' students must publish in order to graduate, graduates must already have publications to enter some PhD programmes, and career events like promotions and salary increases for many researchers are tied to number and impact of publications. Publication has also been made easier by new business models that exploit the author-pays model of open access publishing by selling publication opportunities to authors, often with reduced peer review (Hanson et al, 2024; Weingart, 2025). The opportunities for low-quality journal publishing as well as the introduction of preprint servers and other internet platforms for publishing without peer review all contribute to a rapid rise of numbers of publications that have not been thoroughly assessed by peers.

The rapid growth of the numbers of publications is accompanied by an increasing contamination of the communication system of science worlds with publications that do not communicate scientific novelties. The number of fraudulent publications and publications produced by computers or paper mills is rising faster than the overall number of publications (Richardson et al, 2025). The weakening of peer review and the contamination of the publication system undermine the role of distributive intermediaries, which in turn endangers the trust of community members in publications whose authors they don't know, which so far has been one of the foundations of joint knowledge production by scientific communities. A likely response is that members of scientific communities exclude unknown authors from their screening of the community's literature.

The consequences of these changes since the Second World War for authority relations in science worlds can be summarized as follows. First, the number of practitioner-led and intermediary-led science worlds in which practitioner communities share epistemic authority with audiences and intermediaries has been increasing and is likely to further increase as the dependence of societies on science keeps growing. Second, the diversity of intermediary interests and authority relationships is likely to grow due to both the diversification of funding sources for science and to an increasing interest of audiences of health-related and climate-related science worlds in scientific novelties. This does not necessarily imply a diminishing authority of practitioners, who can retain their authority in science worlds where they gain authority over the distribution of competitive grant funding. Third, in practitioner-led and practitioner-dominated science worlds, the inflationary growth of the number of publications, and among them the proportion of untrustworthy publications, is likely to force the separation of cores of practitioner communities from their peripheries. With this development, members of scientific communities who execute intermediary roles in the assessment and distribution of publications will be able to retain their authority and thus the practitioner community's authority over standards for the assessment of novelties. Fourth, techno-nationalism might split science worlds by intermediaries forcing practitioner communities to split along political lines.

The impact of these changes for innovations in future science worlds is difficult to predict. Generally, the increasing diversity of intermediaries in science worlds increases the likelihood for practitioners to gain access to intermediaries that support specific innovations. Since science worlds appear to adapt to changing socio-economic conditions by practitioner communities excluding members from their novelty production, it is likely that creating and developing innovations could turn into a privilege of a shrinking, well equipped and well-connected elite. This focus of practitioner communities' elites on themselves is likely to reduce the epistemic diversity of contributions from which innovations can be developed. Nevertheless, opportunities to develop the different kinds of innovations discussed in this book will remain.

This somewhat optimistic outlook is of course contingent on the continuing willingness of the state to fund research whose current utility is not visible and which may be used to undermine its policies or ideological positions. As recent developments in several countries have highlighted, this aspect of the societal contract between science and society may be suspended temporarily or permanently. Since practitioner communities in the sciences are international and produce novelties under diverse socio-economic conditions and authority relations, sudden interventions in some countries are unlikely to endanger the novelty production of a whole community. However, if national priorities and governance structures align to the same ideology (as had happened 80 years ago in the wake of Vannevar Bush's 'endless frontier' argument), all science worlds may become intermediary dominated, and the opportunities for scientific innovation may be drastically altered in many of them as a result.

The analysis of lasting change in novelty-driven production in the modern arts and sciences

In this book we have developed and applied a comparative framework for analysing the causes and consequences of lasting change in the modern arts and sciences, partly in response to the observation that no general theoretical accounts have emerged from numerous case studies of such change processes. Seminal proposals like Thomas Kuhn's model of scientific change as the replacement of paradigms captured aspects of lasting change in some fields of the sciences but cannot explain how lasting change is produced in fields of cultural production under changing socio-economic conditions. Economic theories of innovation are of little help for answering this question because they have been developed for 'humdrum industries' (Caves, 2000) and thus cannot answer the question of how lasting change can be conceptualized and analysed in fields of cultural production in which the production of novelties is the dominant purpose of practitioners' creative activity.

The main argument we developed in this book is that the study of this kind of change benefits from a systematic comparison of the modern arts and sciences as fields of novelty-driven production in which creative practitioners must respond to the novelty production of others if they want to make legitimate contributions. We developed this argument in four steps. First, we traced the institutionalization of the modern arts and sciences as novelty-driven fields by reconstructing major societal changes and changes in creative practices between the 16th and 19th centuries. For the comparative analysis of the modern arts and sciences, we developed Becker's concept of art worlds and similar suggestions in the sociology of science in combination with Bourdieu's conceptualizations of roles and authority structures in artistic and scientific fields. In this approach, these institutionalization processes established modern art worlds and science worlds that consisted of *practitioner communities* whose members addressed shared aesthetic/epistemic problems and thus had to respond to each other's novelty production, *audiences* for these novelties and *intermediaries* mediating between audiences and practitioner communities and enabling the production and appreciation of novelties.

In a second step, we developed a concept of innovations as processes that change the purposes and practices of creative practitioners through the continuing development of novelty production. This conceptualization responds to weaknesses of the traditional approach to innovation which have already been pointed out by Nathaniel Rosenberg and which become particularly problematic when applied to fields of creative production. The concept of innovation applied here emphasizes the processual nature of innovation as constantly changing the production of novelties and distinguishes kinds of innovations according to the nature of discontinuous change and the level of collective commitment to changes in the producer community. Combining these two dimensions enabled us to distinguish *drift*, *process*, *complementary* and *displacement* innovations, and showed how the processes bringing about different kinds of lasting change in the arts and sciences can be understood in these terms.

We then identified the structures of art worlds and science worlds that are likely to affect the development of different kinds of innovations. They include productive relationships between members of practitioner communities and the corresponding distribution of authority both within these communities and between them, their audiences and intermediaries. Three types of art worlds with different levels of practitioner authority – *intermediary-led, practitioner-led and practitioner-dominated* – and with *integrated* or *polycentric* authority structures of practitioner communities can be distinguished. The different authority structures in these art worlds and science worlds can be expected to affect decisions by creative practitioners to develop an innovation and their opportunities to do so.

In a final step, we used the empirical literature to construct summary case studies of aesthetic and epistemic innovations in different kinds of art worlds and science worlds. Despite the uneven distribution of this literatures' attention between kinds of innovations, art worlds and science worlds, it was possible to demonstrate how shifts in the distribution of authority between incumbent and new intermediaries and new audiences facilitated major innovations in intermediary-led art worlds and the alignment of intermediary and practitioner interests concerning particular innovations as well as how the approval by growing audiences supported these innovations in practitioner-led art and science worlds. They also confirmed that the development of innovations in practitioner-dominated science worlds still depended on audience and intermediary support for practitioner-dominated science. The case studies not only provided a first theoretical account of conditions facilitating and hindering lasting change in novelty-driven fields of cultural production but also suggested some hypotheses about the impact of current changes in socio-economic conditions for the arts and sciences on the distribution of authority in art worlds and science worlds, and thus ultimately on the opportunities for practitioners to develop innovations.

Our analysis confirmed the historical contingency of novelty-driven production. The reconstruction of its institutionalization showed that novelty-driven production in both the arts and the sciences occurred under specific socio-economic conditions and is currently changing in response to the continuing evolution of these conditions. The stage of development of cultural production whose impact on lasting change in creative production was investigated in this book is itself a phase in the development of creative production in society. The definition of the current situation as a relatively stable phase abstracts from frequent changes. This is justified because similar and relatively stable social structures in the arts and sciences exist despite the differences between artistic and scientific creative practices.

Specific conditions enabling or facilitating lasting change in this phase could be identified through comparative case studies. While limitations of the empirical material prevented the identification of necessary and sufficient conditions for innovations in novelty driven production, the importance of specific changes in the authority of actors and of different forms of alignment of interests of audiences and intermediaries with interests of practitioners developing an innovation became visible in the comparative analysis. The brief account of some current changes in authority structures of art worlds and science worlds demonstrated that the conditions facilitating innovations are also changing and await further comparative analysis.

The comparative analysis also highlighted the shifting dynamics of innovation and community development and their interactions. The nature of innovations may change in the process of their development, as when process innovations turn into complementary innovations if their development makes it possible for practitioners to address new aesthetic/epistemic problems. The structure of practitioner communities may also change, as when the displacement of a Polycentric community's shared background integrates the community. Identifying such cases depends on the application of a comparative framework that is sensitive to change in the empirical data over time.

The discussion of authority structures in art worlds and science worlds identified a number of imbalances in studies of artistic and scientific innovations and suggested some ways of remedying them. Practitioner communities have received great attention in science studies in the wake of Kuhn's work, while studies of artistic innovations have tended to focus on artistic movements advancing innovations but paid less attention to broader practitioner communities until Cottington's (2022) study of secessionist and avant-garde social formations. In contrast, while intermediaries and audiences have received a considerable attention in studies of the arts, researchers in science studies have investigated them mostly in isolation from practitioner communities, and rarely in their intermediary and audience roles.

The comparative study of different kinds of innovations in different art worlds and science worlds was further supported by concepts that could be applied to conditions for innovations occurring in both the arts and the sciences. First, the analysis demonstrated that the four kinds of innovations distinguished in this book occur in both the arts and the sciences, and that conditions for different kinds of innovations can thus be comparatively analysed. It focused on the processes in which innovations were developed rather than on particular innovation-triggering novelties. Conceptualizing innovations as processes of continuing change of practitioners' purposes and practices avoids the concentration of attention on trigger events and their characteristics.

Second, the introduction of the concept of productive relationships between practitioners enabled a comparative analysis of how and why those in different communities respond in particular ways to different kinds of novelties. Third, the concept of authority relations that was developed by us for comparative studies of scientific innovation could be extended for the comparative analysis of conditions for artistic and scientific innovations. Authority relations within practitioner communities and between practitioner communities' audiences and intermediaries could be empirically identified and compared. Finally, the concept of protected space provided an analytical focus for comparing the opportunities to develop innovations without substantial interference from other members of practitioner communities.

The conceptual contributions developed in this book enabled the comparative analysis of changes in different fields of cultural production to be conducted among both the arts and sciences and across them. They supported comparisons based on secondary analyses of a literature that in most cases addressed problems other than the development of innovations. Thus, while this book did not contribute explicitly to methodological development, it developed, operationalized, applied and thus tested a set of conceptual tools that supported the secondary analysis of a diverse literature on the history and sociology of the arts and sciences.

Considering finally the contributions of this book to future research, the comparative approach to the study of lasting change in the arts and sciences contributed analytical tools and conceptual insights for further theory building. Treating the modern arts and sciences as modes of novelty-driven production made it possible to clearly distinguish them from other forms of production and to highlight their specificity beyond the fact that they involve creative practices. Novelty-driven production occurs when a group of practitioners addresses the same aesthetic/epistemic problems. Practitioners are not just isolated creative producers but respond to each other's creativity. This specific form of creative practice, which the modern arts and sciences have in common, is a historical product and might be replaced by different arrangements. While many empirical studies of such changes have been conducted, their impact on social structures of art worlds and science worlds, on authority relations, or on the opportunities for different kinds of innovations have rarely been considered except for the contribution by Buchholz.

To advance the theoretical understanding of innovations in novelty-driven fields of cultural production and of the current changes of conditions for innovations, systematic comparative studies are necessary that simultaneously consider innovation processes and the dynamics of the social contexts in which they take place. All the secondary analyses on which the empirical part of this book is based depend on a combination of sources because few published case studies of innovations have taken into account the complexity of the social contexts in which they were developed. Our book supports this future work by providing a conceptual design for such case studies.

Advancing our theoretical understanding of recent and current changes in novelty-driven cultural production amounts to a substantial research programme rather than a few projects. However, the book demonstrates that paying attention to the complexities of innovation developments from a comparative perspective will benefit studies of both the arts and the sciences.

We see our exploration as a contribution to a more general account of how novelty-driven production with its specific dynamic and structure develops. The production of novelty is a crucial activity in modern society that enables societies to culturally and cognitively adapt to the rapid change they create. It is a central human activity whose dynamics, evolution and adaptation to changing conditions deserve more attention.

Notes

Chapter 2

1 See also Shiff (1984, pp 125–40) and Cottington (2022, pp 136–42) on the classical inheritance of the avant-garde.

2 This condition may be currently disappearing in some of the arts due to changing socio-economic conditions and changing audiences.

3 There is no term more specific than 'scientific practitioner' (Harris, 2006, p 346) that would match the diversity of people engaged in the production of knowledge prior to the scientific revolution (Findlen, 2006a; Park and Daston, 2006; Schiebinger, 2006; Shapin, 2006).

4 According to Aristotle, 'the four causes – matter, form, efficient cause, and final cause – are the four kinds of causes whose combined presence or action is required for the coming-to-be of a substance and whose combined presence or action best explains the motions or changes that a substance undergoes' (Joy, 2006, pp 73–4).

5 See Kuhn (1977) on the 'Baconian sciences'.

6 See Blair (2006) on natural philosophy, Findlen (2006b) on natural history, Donahue (2006) and Mancosu (2006) on mathematical sciences, and Bennett (2006) on mechanical arts. See also Eamon (2006) and Cooper (2006) on the diversity of places of natural inquiry, Bennett (2006) on mechanical arts and instrument making, Findlen (2006b, pp 465–6) on the collaboration between naturalists and instrument makers and Smith (2006) on the institutionalization of laboratories.

7 Bazerman selected all articles in volumes 1, 5, 10, 15, 20, 25, 30, 35, 40, 50, 60, 70, 80 and 90 of the *Philosophical Transactions* (Bazerman, 1988, p 63).

8 The growth of the university landscape was interrupted by the French Revolution and Napoleon's conquests, which led to a drop in the number of European universities from 143 in 1789 to 83 in 1815. The recovery from this blow took most of the 19th century (Rüegg, 2004, p 3).

Chapter 4

1 The idea of a protected space has been applied to sciences in general being shielded from consequences of error and failure (Krohn and Weyer, 1994) and to research processes (see Hackett, 2005, on a 'protected sphere' for scientific innovations and Luukkonen and Thomas, 2016, on 'negotiated space'). Our use of the concept emphasizes the shielding from interferences by all authoritative agencies including practitioner communities and other actors. The use of two measurable dimensions (time horizon and resources) turns 'protected space' from a metaphor into a variable that enables the comparative analysis of opportunities to build specific protected space under varying conditions.

2 This focus on the positioning use of novelties in productive relationships is distinct from Bourdieu's (1993, pp 61–73) focus on position-taking in authority relationships in art and science worlds, which abstracts from the way in which practitioners deal with aesthetic/epistemic problems.

3 This difference in community integration strongly depends on how a community is delineated in the fluid and overlapping patterns of novelty-driven production (see Chapter 2). *Ceteris paribus*, the more numerous and varied are the aesthetic/epistemic problems and approaches that are pursued in a community, the higher is the likelihood of a community being polycentric.

Chapter 5

1 This section is based on a comparative case study of the development of BEC in the Netherlands, Germany, Switzerland and Spain (Laudel et al, 2014b).

2 German university professors who can apply for professorships at other universities and are offered a position can negotiate 'loyalty packages' with their current university if the university wants them to stay. Loyalty packages are akin to start-up packages but are paid as a reward for not leaving the university.

3 Project funding was provided by external funding agencies: the Deutsche Forschungsgemeinschaft in Germany (DFG), Stichting voor Fundamenteel Onderzoek der Materie (FOM) in the Netherlands, Swiss National Science Foundation (SNF) in Switzerland and private foundations in Sweden.

4 This section is based on a comparative case study of the development of Evo-devo in the Netherlands, Germany and Switzerland (Laudel et al, 2014a).

References

Abbate, C. and Parker, R. (2012) *A History of Opera: The Last 400 Years*, Allen Lane.

Ackerman, J. (1969) 'The Demise of the Avant Garde', *Comparative Studies in Society and History*, 11(4): 371–84.

Acord, S. (2010) 'Beyond the Head: The Practical Work of Curating Contemporary Art', *Qualitative Sociology*, 33: 447–67.

Aljets, E. (2014) *Der Aufstieg der empirischen Bildungsforschung: Ein Beitrag zur institutionalistischen Wissenschaftssoziologie*, Springer.

Alpers, S. (1983) *The Art of Describing*, University of Chicago Press.

Anstey, P.R. (2005) 'Experimental versus Speculative Natural Philosophy', in P.R. Anstey and J.A. Schuster (eds) *The Science of Nature in the Seventeenth Century: Patterns of Change in Early Modern Natural Philosophy*, Springer, pp 215–42.

Antliff, M. and Leighten, P. (2001) *Cubism and Culture*, Thames and Hudson.

Bacon, F. (1620 [2000]) *The New Organon*, Cambridge University Press.

Baguna, J. and Garcia-Fernandez, J. (2003) 'Evo-Devo: The Long and Winding Road', *The International Journal of Developmental Biology*, 47(7–8): 705–13.

Baumann, S. (2007) *Hollywood Highbrow*, Princeton University Press.

Baxandall, M. (1972) *Painting and Experience in Fifteenth-Century Italy*, Oxford University Press.

Baxandall, M. (1985) *Patterns of Intention: On the Historical Explanation of Pictures*, Yale University Press.

Bazerman, C. (1988) *Shaping Written Knowledge: The Genre and Activity of the Experimental Article in Science*, The University of Wisconsin Press.

Becker, H.S. (1982) *Art Worlds*, University of California Press.

Becker, H.S. and Pessin, A. (2006) 'A Dialogue on the Ideas of "World" and "Field"', *Sociological Forum*, 21(2): 275–86.

Bennett, J.A. (1986) 'The Mechanics' Philosophy and the Mechanical Philosophy', *History of Science*, 24(1): 1–28.

Bennett, J. (2006) 'The Mechanical Arts', in K. Park and L. Daston (eds) *The Cambridge History of Science, Volume 3: Early Modern Science*, Cambridge University Press, pp 673–95.

Berger, K. (2000) *A Theory of Art*, Oxford University Press.
Berlinguette, C.P., Chiang, Y.M., Munday, J.N., Schenkel, T., Fork, D.K., Koningstein et al (2019) 'Revisiting the Cold Case of Cold Fusion', *Nature*, 570(7759): 45–51.
Berman, E.P. (2012) *Creating the Market University: How Academic Science Became an Economic Engine*, Princeton University Press.
Bernard, J.W. (2003) 'Minimalism, Postminimalism, and the Resurgence of Tonality in Recent American Music Author(s)', *American Music*, 21(1): 112–33.
Biagioli, M. (2002) 'From Book Censorship to Academic Peer Review', *Emergences*, 12(1): 11–45.
Blair, A. (2006) 'Natural Philosophy', in K. Park and L. Daston (eds) *The Cambridge History of Science, Volume 3: Early Modern Science*, Cambridge University Press, pp 365–406.
Blumer, H. (1969) 'Fashion: From Class Differentiation to Collective Selection', *The Sociological Quarterly*, 10(3): 275–91.
Bock-Weiss, C. (1981) *Henri Matisse and Neo-Impressionism, 1898–1908*, University of Michigan Research Press.
Boime, A. (1971) *The Academy and French Painting in the Nineteenth Century*, Phaidon Press.
Bonds, M.E. (1996) *After Beethoven: Imperatives of Originality in the Symphony*, Harvard University Press.
Bornmann, L. and Mutz, R. (2015) 'Growth Rates of Modern Science: A Bibliometric Analysis Based on the Number of Publications and Cited References', *Journal of the Association for Information Science and Technology*, 66(11): 2215–22.
Bourdieu, P. (1975) 'The Specificity of the Scientific Field and the Social Conditions of the Progress of Reason', *Social Science Information*, 14: 119–47.
Bourdieu, P. (1993) 'The Market of Symbolic Goods', in P. Bourdieu (ed) *The Field of Cultural Production: Essays on Art and Literature*, Columbia University Press, pp 112–41.
Bourdieu, P. (2004) 'A World Apart', in P. Bourdieu (ed) *Science of Science and Reflexivity*, University of Chicago Press, pp 32–84.
Bourdieu, P. (2017) *Manet: A Symbolic Revolution*, Polity Press.
Bourdieu, P. and Wacquant, L. (1992) *An Invitation to Reflexive Sociology*, University of Chicago Press.
Braczyk, H.J., Cooke, P. and Heidenreich, M. (eds) (2003) *Regional Innovation Systems: The Role of Governances in a Globalized World*, Routledge.
Braun, D. (1993) 'Who Governs Intermediary Agencies? Principal-Agent Relations in Research Policy-Making', *Journal of Public Policy*, 13(2): 135–62.
Brautigan, R. (1967) *Trout Fishing in America*, Four Seasons.

Britz, J.J. and Ponelis, S. (2012) 'Social Justice and the International Flow of Knowledge with Specific Reference to African Scholars', *Aslib Proceedings* 64(5): 462–77.

Brown, K. (2023) 'Art Markets, Epistemic Authority, and the Institutional Curation of Knowledge', *Cultural Studies*, 37(4): 626–46.

Brush, K. (1996) *The Shaping of Art History*, Cambridge University Press.

Buchholz, L. (2022) *The Global Rules of Art: The Emergence and Divisions of a Cultural World Economy*, Princeton University Press.

Burkholder, J.P. (1994) 'The Uses of Existing Music', *Notes*, 50: 851–70.

Büttner, J., Damerow, P., Renn, J., Schemmel, M. and Valleriani, M. (2002) 'Galileo and the Shared Knowledge of His Time', *Max Planck Institute for the History of Science*, [Preprint], nd, Available from: https://www.mpiwg-berlin.mpg.de/sites/default/files/Preprints/P228.pdf [Accessed 19 January 2026].

Bystryn, M. (1978) 'Art Galleries as Gate Keepers: The Case of the Abstract Expressionists', *Social Research: An International Quarterly*, 45: 390–408.

Cairns, J. (1978) *Cancer: Science and Society*, W.H. Freeman and Company.

Cambrosio, A. and Keating, P. (1995) *Exquisite Specificity: The Monoclonal Antibody Revolution*, Oxford University Press.

Capano, G. (2011) 'Government Continues to Do Its Job: A Comparative Study of Governance Shifts in the Higher Education Sector', *Public Administration*, 89(4): 1622–42.

Carroll, N. (1998) *A Philosophy of Mass Art*, Oxford University Press.

Caves, R. (2000) *Creative Industries*, Harvard University Press.

Chalmers, A. (1993) 'The Lack of Excellency of Boyle's Mechanical Philosophy', *Studies in History and Philosophy of Science*, 24(4): 541–64.

Clark, P.P. (1987) *Literary France: The Making of a Culture*, University of California Press.

Clark, W. (2006) *Academic Charisma and the Origins of the Research University*, University of Chicago Press.

Clarke, A.E. (1990) 'A Social Worlds Research Adventure: The Case of Reproductive Science', in S.E. Cozzens and T.F. Gieryn (eds) *Theories of Science in Society*, Indiana University Press, pp 15–42.

Clignet, R. (1985) *The Structure of Artistic Revolutions*, The University of Pennsylvania Press.

Cohen, J.B. (1985) *Revolution in Science*, The Belknap Press of Harvard University Press.

Cohen, H.F. (1994) *The Scientific Revolution: A Historiographical Inquiry*, The University of Chicago Press.

Collins, R. (1998) *The Sociology of Philosophies. A Global Theory of Intellectual Change*, Cambridge University Press.

Cooper, A. (2006) 'Homes and Households', in K. Park and L. Daston (eds) *The Cambridge History of Science, Volume 3: Early Modern Science*, Cambridge University Press, pp 224–37.
Coover, R. (1968) *The Universal Baseball Association, Inc., J Henry Waugh Prop.*, Random House.
Coover, R. (1969) *Pricksongs & Descants*, Dutton.
Coover, R. (1977) *The Public Burning*, Viking.
Cottington, D. (1998) *Cubism in the Shadow of War: The Avant-Garde and Politics in Paris 1905–1914*, Yale University Press.
Cottington, D. (2004) *Cubism and Its Histories*, Manchester University Press.
Cottington, D. (2022) *Radical Art and the Formation of the Avant-Garde*, Yale University Press.
Cousins, M. (2011) *The Story of Film* (2nd edn), Pavilion Books.
Cox, N. (2000) *Cubism*, Phaidon.
Cozzens, S.E. (1989) *Social Control and Multiple Discovery in Science: The Opiate Receptor Case*, SUNY Press.
Crane, D. (1972) *Invisible Colleges: Diffusion of Knowledge in Scientific communities*, The University of Chicago Press.
Crane, D. (1976) 'Reward Systems in Art, Science, and Religion', *American Behavioral Scientist*, 19(6): 719–34.
Crisp, C. (1993) *The Classic French Cinema 1930–1960*, Indiana University Press.
Crossthwaite, P. (2019) *The Market Logics of Contemporary Fiction*, Cambridge University Press.
Crow, T. (1985) *Painters and Public Life in Eighteen Century Paris*, Yale University Press.
Currie, M. (1995) 'Introduction', in M. Currie (ed) *Metafiction*, Longman, pp 1–18.
Daniels, M. (2019) 'Restricting the Transnational Movement of "Knowledgeable Bodies": The Interplay of US Visa Restrictions and Export Controls in the Cold War', in J. Krige (ed) *How Knowledge Moves: Writing the Transnational History of Science and Technology*, The University of Chicago Press, pp 35–61.
Daston, L. (1991) 'The Ideal and Reality of the Republic of Letters in the Enlightenment', *Science in Context*, 4(2): 367–86.
Daston, L. (1998a) 'The Nature of Nature in Early Modern Europe', *Configurations*, 6(2): 149–72.
Daston, L. (1998b) 'The Language of Strange Facts in Early Modern Science', in T. Lenoir (ed) *Inscribing Science. Scientific Texts and the Materiality of Communication*, Stanford University Press, pp 20–38.
Dear, P. (1985) 'Totius in Verba: Rhetoric and Authority in the Early Royal Society', *ISIS*, 76(2): 145–61.

Dear, P. (1987) 'Jesuit Mathematical Science and the Reconstitution of Experience in the Early Seventeenth Century', *Studies in History and Philosophy of Science*, 18(2): 133–57.

Dear, P. (1990) 'Miracles, Experiments, and the Ordinary Course of Nature', *ISIS*, 81(4): 663–83.

Dear, P. (1991) 'Narratives, Anecdotes, and Experiments: Turning Experience into Science in the Seventeenth Century', in *The Literary Structure of Scientific Argument: Historical Studies*, University of Pennsylvania Press, pp 135–63.

Dear, P. (1995) *Discipline & Experience: The Mathematical Way in the Scientific Revolution*, University of Chicago Press.

Dear, P. (2005) 'What Is the History of Science the History Of? Early Modern Roots of the Ideology of Modern Science', *ISIS*, 96(3): 390–406.

Dear, P. (2006) 'The Meanings of Experience', in K. Park and L. Daston (eds) *The Cambridge History of Science, Volume 3: Early Modern Science*, Cambridge University Press, pp 106–31.

Dedehayir, O. and Steinert, M. (2016) 'The Hype Cycle Model: A Review and Future Directions', *Technological Forecasting and Social Change*, 108: 28–41.

Denzin, N.K. (2009) 'The Elephant in the Living Room: Or Extending the Conversation about the Politics of Evidence', *Qualitative Research*, 9(2): 139–60.

DeVeaux, S. (1997) *The Birth of Bebop: A Social and Musical History*, University of California Press.

DiMaggio, P.J. and Powell, W.W. (1983) 'The Iron Cage Revisited: Institutional Isomorphism and Collective Rationality in Organizational Fields', *American Sociological Review*, 48(2): 147–60.

Donahue, W. (2006) 'Astronomy', in K. Park and L. Daston (eds) *The Cambridge History of Science: Volume 3 Early Modern Science*, Cambridge University Press, pp 562–95.

Drori, G.S., Meyer, J.W., Ramirez, F.O. and Schofer, E. (2003) *Science in the Modern World Polity: Institutionalization and Globalization*, Stanford University Press.

Duffy, A. (2024) 'Weather or Not: A Brief Introduction to Climate History', *Agricultural History*, 98(3): 413–26.

Eamon, W. (1984) 'Arcana Disclosed: The Advent of Printing, the Books of Secrets Tradition and the Development of Experimental Science in the Sixteenth Century', *History of Science*, 22(2): 111–50.

Eamon, W. (1985) 'From the Secrets of Nature to Public Knowledge: The Origins of the Concept of Openness in Science', *Minerva*, 23: 321–47.

Eamon, W. (1991) 'Court, Academy, and Printing House: Patronage and Scientific Career in Late Renaissance Italy', in B.T. Moran (ed) *Patronage and Institutions: Science, Technology and Medicine at the European Court*, The Boydell Press, pp 25–50.

Eamon, W. (2006) 'Markets, Piazzas, and Villages', in K. Park and L. Daston (eds) *The Cambridge History of Science, Volume 3: Early Modern Science*, Cambridge University Press, pp 206–23.

Edge, D. and Mulkay, M.J. (1976) *Astronomy Transformed: The Emergence of Radio Astronomy in Britain*, John Wiley & Sons.

Erlich, V. (1973) 'Russian Formalism', *Journal of the History of Ideas*, 34(4): 627–38

Elbers, A. (2012) 'The Establishment of the New Field of Radio Astronomy in the Post-War Netherlands: A Search for Allies and Funding', *Centaurus*, 54(4): 265–85.

Elias, N. (1993) *Mozart: Portrait of a Genius*, Polity Press.

Everdell, W. (1997) *The First Moderns*, University of Chicago Press.

Fallani, L. and Kastberg, A. (2015) 'Cold Atoms: A Field Enabled by Light', *EPL – Europhysics Letters*, 110(5): 1–7.

Febvre, L. and Martin, H.J. (1976) *The Coming of the Book*, Verso.

Federman, R. (1971) *Double or Nothing*, Swallow Press.

Feingold, M. (2016) '"Experimental Philosophy": Invention and Rebirth of a Seventeenth-Century Concept', *Early Science and Medicine*, 21(1): 1–28.

Fenn, J. and Raskino, M. (2008) *Mastering the Hype Cycle: How to Choose the Right Innovation at the Right Time*, Harvard Business School Press.

Findlen, P. (2006a) 'Anatomy Theaters, Botanical Gardens, and Natural History Collections', in K. Park and L. Daston (eds) *The Cambridge History of Science, Volume 3: Early Modern Science*, Cambridge University Press, pp 272–91.

Findlen, P. (2006b) 'Natural History', in K. Park and L. Daston (eds) *The Cambridge History of Science, Volume 3: Early Modern Science*, Cambridge University Press, pp 435–68.

Fitzgerald, M. (1995) *Making Modernism: Picasso and the Making of the Market for Twentieth Century Art*, University of California Press.

Fleck, L. (1935) *Entstehung und Entwicklung einer wissenschaftlichen Tatsache: Einführung in die Lehre vom Denkstil und Denkkollektiv*, Benno Schwabe & Co.

Fligstein, N. and McAdam, D. (2012) *A Theory of Fields*, Oxford University Press.

Floridi, L. (2002) 'What Is the Philosophy of Information?', *METAPHILOSOPHY*, 33(1/2): 123–45.

Fontes da Costa, P. (2002) 'The Making of Extraordinary Facts: Authentication of Singularities of Nature at the Royal Society of London in the First Half of the Eighteenth Century', *Studies in History and Philosophy of Science*, 33: 265–88.

Fortunato, S., Bergstrom, C.T., Börner, K., Evans, J.A., Helbing, D., Milojević et al (2018) 'Science of Science', *Science*, 359(6379): 1–7.

Fourcade, M. and Khurana, R. (2013) 'From Social Control to Financial Economics', *Theory and Society*, 42(2): 121–59.

Fraser, L. (2017) 'The Barbizon School (1830–1870): Expanding the Landscape of the Modern Art Market', *The Arbutus Review*, 8(1): 4–14.

Freire, L.O. and de Andrade, D.A. (2021) 'Preliminary Survey on Cold Fusion: It's Not Pathological Science and May Require Revision of Nuclear Theory', *Journal of Electroanalytical Chemistry*, 903: 1–10.

Friend, I. (1973) 'Mythodology in Finance', *The Journal of Finance*, 28(2): 257–72.

Friend, I. (1977) 'Recent Developments in Finance', *Journal of Banking and Finance*, 1(2): 103–17.

Friend, I., Westerfield, R. and Granito, M. (1978) 'New Evidence on the Capital Asset Pricing Model', *The Journal of Finance*, 33(3): 903–21.

Fry, E.F. (1966) *Cubism*, Thames and Hudson.

Fry, E.F. (1988) 'Picasso, Cubism, and Reflexivity', *Art Journal*, 47(4): 296–310.

Fujimura, J.H. (1987) 'Constructing "Do-able" Problems in Cancer Research: Articulating Alignment', *Social Studies of Science*, 17(2): 257–93.

Fujimura, J.H. (1988) 'The Molecular Biological Bandwagon in Cancer Research: Where Social Worlds Meet', *Social Problems*, 35(3): 261–83.

Fujimura, J.H. (1996) *Crafting Science: A Sociohistory of the Quest for the Genetics of Cancer*, Harvard University Press.

Galenson, D.W. (2006) *Artistic Capital*, Routledge.

Galenson, D.W. and Jenssen, R. (2002) 'Careers and Canvases: The Rise of the Market for Modern Art in the Nineteenth Century', Working paper 9123, National Bureau of Economic Research.

Galison, P. and Stump, D.J. (1996) *The Disunity of Science: Boundaries, Contexts, and Power*, Stanford University Press.

Garber, D. (2006) 'Physics and Foundations', in K. Park and L. Daston (eds) *The Cambridge History of Science, Volume 3: Early Modern Science*, Cambridge University Press, pp 21–69.

George, W. (1980) 'Sources and Background to Discoveries of New Animals in the Sixteenth and Seventeenth Centuries', *History of Science*, 18(2): 79–104.

Georgescu, I. (2020) '25 years of BEC', *Nature Reviews Physics*, 2: 396.

Gerbod, P. (2004a) 'Relations with Authority', in W. Rüegg (ed) *A History of the University in Europe: Volume III: Universities in the Nineteenth and early Twentieth Centuries (1800–1945)*, Cambridge University Press, pp 83–100.

Gerbod, P. (2004b) 'Resources and Management', in W. Rüegg (ed) *A History of the University in Europe: Volume III: Universities in the Nineteenth and early Twentieth Centuries (1800–1945)*, Cambridge University Press, pp 101–21.

Gerson, E.M. (1983) 'Scientific Work and Social Worlds', *Knowledge: Creation, Diffusion, Utilization*, 4(3): 357–77.

Gilbert, G.N. (1976) 'The Transformation of Research Findings into Scientific Knowledge', *Social Studies of Science*, 6(3/4): 281–306.

Gilmore, S. (1987) 'Coordination and Convention: The organization of the concert world', *Symbolic Interaction*, 10(2): 209–27.

Gilmore, S. (1988) 'Schools of Activity and Innovation', *The Sociological Quarterly*, 29(2): 203–19.

Gioia, T. (2011) *The History of Jazz*, Oxford University Press.

Gippel, J.K. (2012) 'A Revolution in Finance?', *Australian Journal of Management*, 38(1): 125–46.

Giudice, G.F., Sibiryakov, S. and Strumia, A. (2012) 'Interpreting OPERA Results on Superluminal Neutrino', *Nuclear Physics B*, 861(1): 1–16.

Gläser, J. (2006) *Wissenschaftliche Produktionsgemeinschaften. Die soziale Ordnung der Forschung*. Campus.

Gläser, J. and Schimank, U. (2014) ‚Autonomie als Resistenz gegen Beeinflussung. Forschungshandeln im organisatorischen und politischen Kontext', in M. Franzen, A. Jung, D. Kaldewey and J. Korte (eds) *Autonomie revisited: Beiträge zu einem umstrittenen Grundbegriff in Wissenschaft, Kunst und Politik*, Beltz Juventa, pp 41–61.

Gläser, J. and Laudel, G. (2016) 'Governing Science: How Science Policy Shapes Research Content', *European Journal of Sociology/Archives Européennes de Sociologie*, 57(1): 117–68.

Gläser, J., Laudel, G., and Whitley, R. (2026) 'Innovations in the Sciences, Social Sciences, and Humanities', in B. Blättel-Mink, I. Schulz-Schaeffer and A. Windeler (eds) *Handbook Innovation Research*, Springer, in press.

Gläser, J., Aljets, E., Gorga, A., Hedmo, T., Håkansson, E. and Laudel, G. (2014) 'Path Dependence and Policy Steering in the Social Sciences: The Varied Impact of International Large Scale Student Assessment on the Educational Sciences in Four European Countries', in R. Whitley and J. Gläser (eds) *Organizational Transformation and Scientific Change: The Impact of Institutional Restructuring on Universities and Intellectual Innovation*, Emerald Group, pp 267–95.

Gnyp, M. (2020) *The Shift: Art and the Rise to Power of Contemporary Collectors*, Skira.

Godin, B. (2012) '"Innovation Studies": The Invention of a Speciality', *Minerva*, 50: 397–421.

Godin, B. (2016) 'Technological Innovation: On the Origin and Development of an Inclusive Concept', *Technology and Culture*, 57(3): 527–56.

Goehr, L. (1992) *The Imaginary Museum of Musical Works*, Oxford University Press.

Golding, J. (1968) *Cubism: A History and an Analysis 1907–1914*, Faber and Faber.

Golinski, J. (1998) *Making Natural Knowledge: Constructivism and the History of Science*, Cambridge University Press.

Golinski, J. (2003) 'Chemistry', in R. Porter (ed) *The Cambridge History of Science: Volume 4 Eighteenth-Century Science*, Cambridge University Press, pp 375–96.

Gombrich, E. (1966) 'The Renaissance Conception of Artistic Progress and its Consequences', in E. Gombrich (ed) *Norm and Form: Studies in the Art of the Renaissance*, Phaidon, pp 1–10.

Gomery, D. (1996) 'The Hollywood Studio System', in G. Nowell-Smith (ed) *The Oxford History of World Cinema*, Oxford University Press, pp 43–53.

Gordon, R.A. and Howell, J.E. (1959) *Higher Education for Business*, Columbia University Press.

Graham, P. (1996) 'New Directions in French Cinema', in G. Nowell-Smith (ed) *The Oxford History of World Cinema*, Oxford University Press, pp 576–86.

Grieser, C. (2018) 'The Influence of Companies on Scientific Communities. A Qualitative Study of German Automotive Engineering', [Master Thesis] 6 March, Available from: https://www.researchgate.net/publication/325538304_The_Influence_of_Companies_on_Scientific_Communities_A_Qualitative_Study_of_German_Automotive_Engineering/citations [Accessed 20 January 2026].

Gronow, J. (2009) 'Fads, Fashions and "real" Innovations', in E. Shove, F. Trentmann and R. Wilk (eds) *Time, Consumption and Everyday Life: Practice, Materiality and Culture*, Routledge, pp 129–42.

Guston, D.H. (2000) 'The Expanding Role of Peer Review Processes in the United States', in P. Shapira and S. Kuhlmann (eds) *Learning from Science and Technology Policy Evaluation: Experiences from the United States and Europe*, Elgaronline, pp 81–97.

Hackett, E.J. (2005) 'Essential Tensions: Identity, Control, and Risk in Research', *Social Studies of Science*, 35(5): 787–826.

Hahn, R. (1971) *The Anatomy of a Scientific Institution: The Paris Academy of Sciences, 1666–1803*, University of California Press.

Hahn, R. (1975) 'Scientific Research as an Occupation in Eighteenth-Century Paris', *Minerva*, 13(4): 501–13.

Hall, A.R. (1963) 'Merton Revisited, or Science and Society in the Seventeenth Century', *History of Science*, 2(1): 1–16.

Hall, B.K. (2012) 'Evolutionary Development Biology (Evo-Devo): Past, Present and Future', *Evolution: Education and Outreach*, 5: 184–93.

Hall, M.B. (1991) *Promoting Experimental Learning: Experiment and the Royal Society, 1660–1727*, Cambridge University Press.

Hall, R. (1959) 'The Scholar and the Craftsman in the Scientific Revolution', in M. Clagett (ed) *Critical Problems in the History of Science*, The University of Wisconsin Press, pp 3–23.

Hammersley, M. (2018) *The Radicalisation of Ethnomethodology*, University of Manchester Press.

Hanson, M.A., Barreiro, P.G., Crosetto, P. and Brockington, D. (2024) 'The Strain on Scientific Publishing', *Quantitative Science Studies*, 5(4): 823–43.

Harris, S.J. (2006) 'Networks of Travel, Correspondence, and Exchange', in K. Park and L. Daston (eds) *The Cambridge History of Science, Volume 3: Early Modern Science*, Cambridge University Press, pp 341–62.

Haskell, F. (1980) *Patrons and Painters*, Yale University Press.

Heinze, T., Shapira, P., Rogers, J.D. and Senker, J.M. (2009) 'Organizational and Institutional Influences on Creativity in Scientific Research', *Research Policy*, 38(4): 610–23.

Heller, J. (1961) *Catch-22*, Simon & Schuster.

Hessels, L.K., Grin, J. and Smits, R.E.H.M. (2011) 'The Effects of a Changing Institutional Environment on Academic Research Practices: Three Cases from Agricultural Science', *Science and Public Policy*, 38(7): 555–68.

Hicks, D. (2012) 'Performance-Based University Research Funding Systems', *Research Policy*, 41(2): 251–61.

Hollingsworth, J.R. (2008) 'Scientific Discoveries: An Institutionalist and Path-Dependent Perspective', in C. Hannaway (ed) *Biomedicine in the Twentieth Century: Practices, Policies, and Politics*, IOS Press, pp 317–53.

Holmes, F.L. (1989) *Eighteenth-Century Chemistry as an Investigative Enterprise*, Office for History of Science and Technology, University of California.

Holmes, F.L. (1991) 'Argument and Narrative in Scientific Writing', in P. Dear (ed) *The Literary Structure of Scientific Argument: Historical Studies*, University of Pennsylvania Press, pp 164–81.

Holmes, F.L. (1996) 'The Communal Context for Etienne-François Geoffroy's "Table des rapports"', *Science in Context*, 9(3): 289–311.

Holmes, R. (2008) *The Age of Wonder*, HarperCollins.

Home, R.W. (2003) 'Mechanics and Experimental Physics', in R. Porter (ed) *The Cambridge History of Science: Volume 4 Eighteenth-Century Science*, Cambridge University Press, pp 354–74.

Horowitz, N. (2011) *Art of the Deal: Contemporary Art in a Global Financial Market*, Princeton University Press.

Hooykaas, R. (1987) 'The Rise of Modern Science: When and Why?', *British Journal of the History of Science*, 20(4): 453–73.

Hutcheon, L. (1980) *Narcissistic Narratives: The Metafictional Paradox*, Wilfrid Laurier University Press.

Iliffe, R. (1992) '"In the Warehouse": Privacy, Property and Priority in the Early Royal Society', *History of Science*, 30(1): 29–68.

Jensen, M.C. (1972) 'Capital Markets: Theory and Evidence', *Bell Journal of Economics and Management Science*, 3(2): 357–98.

Jensen, R. (1994) *Marketing Modernism in Fin-de-Siecle Europe*, Princeton University Press.

Johnson, R., Watkinson, A. and Mabe, M. (2018) 'The STM Report 2018: An Overview of Scientific and Scholarly Publishing', *International Association of Scientific, Technical and Medical Publishers*, [Report] October, Available from: https://stm-assoc.org/document/the-stm-report-2018-an-overview-of-scientific-and-scholarly-publishing/ [Accessed 20 January 2026].

Jovanovic, F. (2008) 'The Construction of the Canonical History of Financial Economics', *History of Political Economy*, 40(2): 213–42.

Joy, L.S. (2006) 'Scientific Explanation from formal causes to Laws of Nature', in K. Park and L. Daston (eds) *The Cambridge History of Science: Volume 3 Early Modern Science*, Cambridge University Press, pp 70–105.

Katz, R. (1984) 'Collective "Problem Solving" in the History of Music', *Journal of the History of Ideas*, 45(3): 361–77.

Kavesh, R.A. (1970) 'The American Finance Association: 1939–1969', *The Journal of Finance*, 25(1): 1–17.

Kearnes, M. and Wienroth, M. (2011) 'Tools of the Trade: UK Research Intermediaries and the Politics of Impacts', *Minerva*, 49(2): 153–74.

Keenan, M. (1970) 'Models of Equity Valuation: The Great Serm Bubble', *The Journal of Finance*, 25(2): 243–73.

Kemp, M. (1997) *Behind the Picture*, Yale University Press.

Kendall, M.G. (1953) 'The Analysis of Economic Time-Series, Part I: Prices', *Journal of the Royal Statistical Society A*, 116(1): 11–34.

Ketterle, W. (2002) 'When Atoms Behave as Waves: Bose-Einstein Condensation and the Atom Laser (Nobel Lecture)', *ChemPhysChem*, 3(9): 736–53.

Klein, U. (1996) 'The Chemical Workshop Tradition and the Experimental Practice: Discontinuities within Continuities', *Science in Context*, 9(3): 251–87.

Klinkowitz, J. (1980) *Literary Disruptions: The making of a Post-contemporary American Fiction* (2nd edn), University of Illinois Press.

Klinkowitz, J. (2017) 'Metafiction', *Oxford Research Encyclopaedia of Literature*, [Encyclopaedia Entry] 26 July, Available from: https://doi.org/10.1093/acrefore/9780190201098.013.546 [Accessed 20 January 2026].

Krige, J. (2019) 'Export Controls as Instruments to Regulate Knowledge Acquisition in a Globalizing Economy', in J. Krige (ed) *How Knowledge Moves: Writing the Transnational History of Science and Technology*, The University of Chicago Press: pp 62–92.

Krohn, W. and Weyer, J. (1994) 'Society as Laboratory: The Social Risks of Experimental Research', *Science and Public Policy*, 21(3): 173–83.

Kubler, G. (1962) *The Shape of Time: Remarks on the History of Things*, Yale University Press.

Kuhn, T.S. (1961) 'The Function of Measurement in Modern Physical Science', *ISIS*, 52(2): 161–93.

Kuhn, T.S. (1970 [1962]) *The Structure of Scientific Revolutions*, The University of Chicago Press.

Kuhn, T.S. (1977) *The Essential Tension*, University of Chicago Press.

Lakatos, I. (1968) 'Criticism and the Methodology of Scientific Research Programmes', *Proceedings of the Aristotelian Society*, 69: 149–86.

Lamont, M. (2009) *How Professors Think: Inside the Curious World of Academic Judgment*, Harvard University Press.

Lamont, M. (2012) 'Toward a Comparative Sociology of Valuation and Evaluation', *Annual Review of Sociology*, 38(1): 201–21.

Laudel, G. and Weyer, E. (2014) 'Where Have All the Scientists Gone? Building Research Profiles at Dutch Universities and Its Consequences for Research, in R. Whitley and J. Gläser (eds) *Organizational Transformation and Scientific Change: The Impact of Institutional Restructuring on Universities and Intellectual Innovation*, Emerald Group, pp 111–40.

Laudel, G., Benninghoff, M., Lettkemann, E. and Håkansson, E. (2014a) 'Highly Adaptable but not Invulnerable: Necessary and Facilitating Conditions for Research in Evolutionary Developmental Biology', in R. Whitley and J. Gläser (eds) *Organizational Transformation and Scientific Change: The Impact of Institutional Restructuring on Universities and Intellectual Innovation*, Emerald Group, pp 235–65.

Laudel, G., Lettkemann, E., Ramuz, R., Wedlin, L. and Woolley, R. (2014b) 'Cold Atoms – Hot Research: High Risks, High Rewards in Five Different Authority Structures', in R. Whitley and J. Gläser (eds) *Organizational Transformation and Scientific Change: The Impact of Institutional Restructuring on Universities and Intellectual Innovation*, Emerald Group, pp 203–34.

Lave, J. and Wenger, E.C. (1991) *Situated Learning: Legitimate Peripheral Participation*, Cambridge University Press.

Lee, J.J. and Li X. (2023) 'Neo-racism, Neo-nationalism, and the Costs for Scientific Competitiveness: The China Initiative in the United States', *The Review of Higher Education*, 46(3): 285–309.

Lee, K. (2018) 'Neoliberal Marketization of Global Contemporary Visual Art Worlds: Changes in Valuations and the Scope of Local and Global Markets', in Alexander V.D., Hägg S., Häyrynen S. and Sevänen E. (eds) *Art and the Challenge of Markets Volume 2*, Palgrave Macmillan, pp 65–97.

Levaux, C. (2020) *We Have Always Been Minimalist*, University of California Press.

Lewenstein, B.V. (1995) 'From Fax to Facts: Communication in the Cold-Fusion Saga', *Social Studies of Science*, 25(3): 403–36.

Leypoldt, G. (2015) 'Shifting Meridians: US Authorship in World Literary Space', *American Literary History*, 27(4): 768–87.

Liberali, P. and Schier, A.F. (2024) 'The Evolution of Developmental Biology Through Conceptual and Technological Revolutions', *Cell*, 187(14): 3461–95.

Lintner, J. (1965) 'The Valuation of Risk Assets and the Selection of Risky Investments in Stock Portfolio and Capital Budgets', *Review of Economics and Statistics*, 47(1): 13–37.

Long, P.O. (1991a) 'Invention, Authorship, "Intellectual Property", and the Origin of Patents: Notes toward a Conceptual History', *Technology and Culture*, 32(4): 846–884.

Long, P.O. (1991b) 'The Openness of Knowledge: An Ideal and Its Context in 16th-Century Writings on Mining and Metallurgy', *Technology and Culture*, 32(2/1): 318–55.

Long, P.O. (1997) 'Power, Patronage, and the Authorship of Ars: From Mechanical Know-how to Mechanical Knowledge in the Last Scribal Age', *ISIS*, 88(1): 1–41.

Long, P.O. (2001) *Openness, Secrecy, Authorship: Technical Arts and the Culture of Knowledge from Antiquity to the Renaissance*, The Johns Hopkins University Press.

Lopes, P. (2002) *The Rise of a Jazz Art World*, Cambridge University Press.

Lorente, J.P. (2011) *The Museums of Contemporary Art: Notion and Development*, Routledge.

Lundvall, B.-A. (ed) (1992) *National Systems of Innovation*, Pinter.

Luukkonen, T. and Thomas, D.A. (2016) 'The "Negotiated Space" of University Researchers' Pursuit of a Research Agenda', *Minerva*, 54(1): 99–127.

MacKenzie, D. (2003) 'An Equation and its Worlds: Bricolage, Exemplars, Disunity and Performativity in Financial Economics', *Social Studies of Science*, 33(6): 831–68.

MacKenzie, D. and Millo, Y. (2003) 'Constructing a Market, Performing Theory: The Historical Sociology of a Financial Derivatives Exchange', *American Journal of Sociology*, 109(1): 107–45.

Malafronte, A. (2009) 'The History of the *Plein Air* Movement', *American Artist*, 73(802): 20–4.

Mancosu, P. (2006) 'Acoustics and Optics', in K. Park and L. Daston (eds) *The Cambridge History of Science, Volume 3: Early Modern Science*, Cambridge University Press, pp 596–631.

Mangee, N.J. (2015) 'A Kuhnian Perspective on Asset Pricing Theory', *Journal of Economic Methodology*, 22(1): 28–45.

Marie, M. (2003) *The French New Wave*, Blackwell.

Markowitz, J. (1952) 'Portfolio Selection', *Journal of Finance*, 7(1): 77–91.

Martin, B.R. (1978) 'Radio Astronomy Revisited: A Reassessment of the Role of Competition and Conflict in the Development of Radio Astronomy', *The Sociological Review*, 26(1): 27–56.

Martin, B.R. (2016) 'Twenty Challenges for Innovation Studies', *Science and Public Policy*, 43(3): 432–50.

Masterman, M. (1970) 'The Nature of a Paradigm', in I. Lakatos and A. Musgrave (eds) *Criticism and the Growth of Knowledge*, Cambridge University Press, pp 59–89.

Matsumoto, K., Shibayama, S., Kang, B. and Igami, M. (2021) 'Introducing a Novelty Indicator for Scientific Research: Validating the Knowledge-Based Combinatorial Approach', *Scientometrics*, 126(8): 6891–915.

McCain, K.W. (2015) 'Mining Full-Text Journal Articles to Assess Obliteration by Incorporation: Herbert A. Simon's Concepts of Bounded Rationality and Satisficing in Economics, Management, and Psychology', *Journal of the Association for Information Science and Technology*, 66(11): 2187–201.

McClellan III, J. (2003) 'Scientific Institutions and the Organization of Science', in R. Porter (ed) *The Cambridge History of Science, Volume 4: Eighteenth-Century Science*, Cambridge University Press, pp 87–106.

McGurl, M. (2009) *The Program Era: Postwar Fiction and the Rise of Creative Writing*, Harvard University Press.

McKeon, M. (1987) *The Origins of the English Novel 1600–1740*, Johns Hopkins University Press.

Menand, L. (2010) *The Marketplace for Ideas*, W.W. Norton.

Meyer, L. (1956) *Emotion and Meaning in Music*, University of Chicago Press.

Meyer, L. (1974) 'Concerning the Sciences, the Arts – AND the Humanities, *Critical Inquiry*, 1(1): 163–217.

Miles, I. (2016) 'Twenty Years of Service Innovation Research' in M. Toivonen (ed) *Service Innovation*, Springer Japan, pp 3–34.

Miller, M.H. and Modigliani, F. (1961) 'Dividend Policy, Growth and the Valuation of Shares', *Journal of Business*, 34(4): 411–33.

Miozzo, M. and Grimshaw, D. (eds) (2006) *Knowledge Intensive Business Services*, Edward Elgar.

Modigliani, F. and Miller, M.H. (1958) 'The Cost of Capital, Corporation Finance and the Theory of Investment', *American Economic Review*, 48(3): 261–97.

Montias, J. (1982) *Artists and Artisans in Delft*, Princeton University Press.

Morange, M. (1997) 'From the Regulatory Vision of Cancer to the Oncogene Paradigm, 1975–1985', *Journal of the History of Biology* 30(1): 1–29.

Morgan, R. (1978) 'Ives and Mahler: Mutual Responses at the End of an Era', *19th Century Music*, 2(1): 72–81.

Morrell, J.B. (1990) 'Professionalisation', in C. Olby, G.N. Cantor, J.R.R. Christie and M.J.S. Hodge (eds) *Companion to the History of Modern Science*, Routledge, pp 980–9.

Moxey, K. (2001) *The Practice of Persuasion: Paradox and Power in Art History*, Cornell University Press.

Mullins, N.C. (1973) 'The Development of Specialties in Social Science: The Case of Ethnomethodology', *Science Studies*, 3(3): 245–73.

Mulkay, M.J. (1975) 'Three Models of Scientific Development', *The Sociological Review*, 23(3): 509–26.

Mulkay, M.J. and Edge, D.O. (1973) 'Cognitive, Technical and Social Factors in the Growth of Radio Astronomy', *Social Science Information*, 12(6): 25–61.

Müller, G.B. (2007) 'Six Memos for Evo-Devo', in M.D. Laubichler and J. Maienschein (eds) *Embryology to Evo–Devo: A History of Developmental Evolution*, MIT Press, pp 499–524.

Munns, D.P.D. (2013) *A Single Sky: How an International Community Forged the Science of Radio Astronomy*, MIT Press.

Myers, D.G. (2006) *The Elephants Teach: Creative Writing since 1880*, University of Chicago Press.

Myers, M. (2013) *Why Jazz Happened*, University of California Press.

Nelson, R. (ed) (1993) *National Systems of Innovation*, Oxford University Press.

Newman, W.R. (2006) From Alchemy to 'Chymistry', in K. Park and L. Daston (eds) *The Cambridge History of Science, Volume 3: Early Modern Science*, Cambridge University Press, pp 497–517.

North, M. (2013) *Novelty: A History of the New*, University of Chicago Press.

Nye, M.J. (1993) 'National Styles? French and English Chemistry in the Nineteenth and Early Twentieth Centuries', *OSIRIS*, 8: 30–49.

Ogilvie, B.W. (2006) *The Science of Describing: Natural History in Renaissance Europe*, The University of Chicago Press.

Ogilvie, B. (2016) 'Correspondence Networks', in B. Lightman (ed) *A Companion to the History of Science*, John Wiley & Sons, pp 358–71.

O'Neill, P. (2012*) The Culture of Curating and the Curating of Cultures*, MIT Press.

Osler, M.J. (2000) 'The Canonical Imperative: Rethinking the Scientific Revolution', in M.J. Osler (ed) *Rethinking the Scientific Revolution*, Cambridge University Press, pp 3–24.

Owens, T. (1995) *Bebop: The Music and Its Players*, Oxford University Press.

Park, K. and Daston, L.J. (1981) 'Unnatural Conceptions: The Study of Monsters in Sixteenth- and Seventeenth-Century France and England', *Past and Present*, 92: 20–54.

Park, K. and Daston, L. (2006) 'The Age of the New', in K. Park and L. Daston (eds) *The Cambridge History of Science, Volume 3: Early Modern Science*, Cambridge University Press, pp 1–17.

Parkinson, D. (2012) *History of Film*, Thames and Hudson.

Paul, C. (ed) (1975) *The First Modern Museums of Art*, Getty Publications.

Peterson, R. and Berger, D.G. (1975) 'Cycles in Symbol Production: The Case of Popular Music', *American Sociological Review*, 40(2): 158–73.

Pierson, F.C., Bach, G.L., Griswold, J.A., Timms, H.L., Ireson, W.G. and Bridgman, D.S. (1959) *The Education of American Businessmen*, McGraw-Hill.

Pinch, T. (1986) *Confronting Nature: The Sociology of Solar Neutrino Detection*, Reidel.

Pinch, T. (1994) 'Cold Fusion and the Sociology of Scientific Knowledge', *Technical Communication Quarterly*, 3(1): 85–100.

Pomian, K. (1987) *Collectors and Curiosities: Paris and Venice, 1500–1800*, Polity Press.

Porter, R. (1996) 'The Scientific Revolution and Universities', in H. de Ridder-Symoens (ed) *Universities in Early Modern Europe, 1500–1800*, Cambridge University Press, pp 531–62.

Porter, R. (2000) *English Society in the Eighteenth Century*, Penguin.

Porter, R. (ed) (2003) *The Cambridge History of Science, Volume 4: Eighteenth-Century Science*, Cambridge University Press.

Prettejohn, E. (1997) 'Aesthetic Value and the Professionalization of Victorian Art Criticism 1837–78', *Journal of Victorian Culture*, 2(1): 71–94.

Principe, L.M. (2013) *The Secrets of Alchemy*, The University of Chicago Press.

Quemin, A. (2013) 'International Contemporary Art Fairs in a "Globalized" Art Market', *European Societies*, 15(2): 162–77.

Quemin, A. (2020) 'The Market and Museums: The Increasing Power of Collectors and Private Galleries in the Contemporary Art World', *Journal of Visual Art Practice*, 19(3): 211–24.

Rachlin, S. (1993) 'Anatomy of a Film Revolution: The Case of the Nouvelle Vague', *Poetics*, 21(5): 429–42.

Richardson, R.A.K., Hong, S.S., Byrne, J.A,. Stoeger, T. and Amaral, L.A.N. (2025) 'The Entities Enabling Scientific Fraud at Scale Are Large, Resilient, and Growing Rapidly', *Proceedings of the National Academy of Sciences*, 122(32): e2420092122.

Rip, A. (1994) 'The Republic of Science in the 1990s', *Higher Education*, 28: 3–32.

Rip, A. (1995) 'New Combinations', *European Review*, 3(1): 83–92.

Rip, A. (2011) 'Protected Spaces of Science: Their Emergence and Further Evolution in a Changing world', in M. Carrier and A. Nordmann (eds) *Science in the context of application*, Springer, pp 197–220.

Rogers, E. (2003) *Diffusion of Innovations* (5th edn), Free Press.

Rosen, C. (2000) *Critical Entertainments*, Harvard University Press.

Rosenau, J.N. (2004) 'Strong Demand, Huge Supply: Governance in an Emerging Epoch', in I. Bache and M. Flinders (eds) *Multi-level Governance*, Oxford Scholarship Online, pp 31–48.

Rosenberg, N. (1975) 'Problems in the Economist's Conceptualisation of Technological Innovation', *History of Political Economy*, 7(4): 456–81.

Ross, A. (2012) *The Rest is Noise: Listening to the twentieth century*, Fourth Estate.

Ross, S.A. (1976) 'The Arbitrage Theory of Capital Asset Pricing', *Journal of Economic Theory*, 13: 341–60.

Ross, S.A. (1978) 'The Current Status of the Capital Asset Pricing Model', *Journal of Finance*, 33(3): 885–901.

Rossi, P. (1970) *Philosophy, Technology and the Arts in the Early Modern Era*, Harper & Row.
Rutherford, T. and Holmes, J. (2008) 'Engineering Networks: University–Industry Networks in Southern Ontario Automotive Industry Clusters', *Cambridge Journal of Regions, Economy and Society*, 1(2): 247–64.
Rüegg, W. (2004) 'Themes', in W. Rüegg (ed) *A History of the University in Europe Volume III: Universities in the Nineteenth and early Twentieth Centuries*, Cambridge University Press, pp 3–31.
Ryan, R.J. (1982) 'Capital Market Theory – A Case Study in Methodological Conflict', *Journal of Business Finance and Accounting*, 9(4): 443–58
Sallis, F. (2018) 'Borrowing in Music', in N. Donin (ed) *The Oxford Handbook of the Creative Process in Music* (online edn) Oxford Academic, https://doi.org/10.1093/oxfordhb/9780190636197.013.19.
Schaffer, S. (1989) 'Glass Works: Newton's Prisms and the Uses of Experiment', in D. Gooding, T. Pinch and S. Schaffer (eds) *The Uses of Experiment*, Cambridge University Press, pp 67–104.
Schatz, T. (1996) 'Hollywood: The Triumph of the Studio System', in G. Nowell-Smith (ed) *The Oxford History of World Cinema*, Oxford University Press, pp 220–35.
Scherer, F.M. (2004) *Quarter Notes and Bank Notes: The Economics of Music Composition in the Eighteenth and Nineteenth Centuries*, Princeton University Press.
Schiebinger, L. (2006) 'Women of Natural Knowledge', in K. Park and L. Daston (eds) *The Cambridge History of Science, Volume 3: Early Modern Science*, Cambridge University Press, pp 192–205.
Schorske, C. (1983) 'Mahler and Ives: Populist Archaism and Musical Innovation', *Bulletin of the American Academy of Arts and Sciences*, 37: 20–39.
Schütz, A. (1962) 'The Problem of Social Reality', in A. Schütz (ed) *Collected Papers volume 1*, Martinus Nijhoff, pp 3–94.
Sgourev, S. (2013) 'How Paris Gave Rise to Cubism (and Picasso): Ambiguity and Fragmentation in Radical Innovation', *Organization Science*, 24(6): 1601–17.
Shapin, S. (1984) 'Pump and Circumstance: Robert Boyle's Literary Technology', *Social Studies of Science*, 14(4): 481–520.
Shapin, S. (1988) 'The House of Experiment in Seventeenth-Century England', *ISIS*, 79(3): 373–404.
Shapin, S. (1994) *A Social History of Truth: Civility and Science in Seventeenth-Century England*, The University of Chicago Press.
Shapin, S. (1996) *The Scientific Revolution*, The University of Chicago Press.
Shapin, S. (2006) 'The Man of Science', in K. Park and L. Daston (eds) *The Cambridge History of Science, Volume 3: Early Modern Science*, Cambridge University Press, pp 179–205.

Shapin, S. and Schaffer, S. (1985) *Leviathan and the Air-Pump: Hobbes, Boyle, and the Experimental Life*, Princeton University Press.

Shapiro, B.J. (2000) *A Culture of Fact: England, 1550–1720*, Cornell University Press.

Shapiro, B.J. (2002) 'Testimony in Seventeenth-Century English Natural Philosophy: Legal Origins and Early Development', *Studies in History and Philosophy of Science*, 33(2): 243–63.

Sharpe, W.F. (1964) 'Capital Asset Prices: A Theory of Market Equilibrium under Conditions of Risk', *Journal of Finance*, 19(3): 425–42.

Shibayama, S. and Wang, J. (2020) 'Measuring Originality in Science', *Scientometrics*, 122(1): 409–27.

Shiff, R. (1984) *Cezanne and the End of Impressionism. A Study of the Theory, Technique, and Critical Evaluation of Modern Art*, The University of Chicago Press.

Shiller, R.J. (2003) 'From Efficient Markets Theory to Behavioral Finance', *The Journal of Economic Perspectives*, 17(1): 83–104.

Shiner, L. (2001) The *Invention of Art: A Cultural History*, University of Chicago Press.

Simmel, G. (1957 [1904]) 'Fashion', *American Journal of Sociology*, 62(6): 541–58.

Simon, B. (1999) 'Undead Science: Making Sense of Cold Fusion after the (Arti)fact', *Social Studies of Science*, 29(1): 61–85.

Simon, B. (2002) *Undead Science: Science Studies and the Afterlife of Cold Fusion*, Rutgers University Press.

Smith, C. (2012) *The Company of Artists: The Origins of the Royal Academy of Arts in London*, Modern Art Press.

Smith, P.H. (2006) 'Laboratories', in K. Park and L. Daston (eds) *The Cambridge History of Science, Volume 3: Early Modern Science*, Cambridge University Press, pp 290–305.

Smith, P.H. (2009) 'Science on the Move: Recent Trends in the History of Early Modern Science', *Renaissance Quarterly*, 62(2): 345–75.

Sommer, R.J. (2009) 'The Future of Evo–Devo: Model Systems and Evolutionary Theory', *Nature Reviews Genetics*, 10(6): 416–22.

Sorrentino, G. (1970) *Steelwork*, Pantheon.

Sorrentino, G. (1971) *Imaginative Qualities of Actual Things*, Pantheon.

Star, S.L. and Gerson, E.M. (1987) 'The Management and Dynamics of Anomalies in Scientific Work', *Sociological Quarterly*, 28(2): 147–69.

Stevenson, R. (2004) *The Last of England?*, Oxford University Press.

Studer, K.E. and Chubin, D.E. (1980) *The Cancer Mission: Social Contexts of Biomedical Research*, Sage.

Stump, R. (1998) 'Place and Innovation in Popular Music: The Bebop Revolution in Jazz', *Journal of Cultural Geography*, 18(1): 11–34.

Sukenik, R. (1968) *Up*, Dial Press.

Summers, D. (1981) *Michelangelo and the Language of Art*, Princeton University Press.

Teece, D.J. (2010) 'Business Models, Business Strategy and Innovation', *Long Range Planning*, 43(2): 172–94.

Tushman, M.L. and Anderson, P. (1986) 'Technological Discontinuities and Organizational Environments', *Administrative Science Quarterly*, 31(3): 439–65.

van Helvoort, T. (1999) 'A Century of Research into the Cause of Cancer: Is the New Oncogene Paradigm Revolutionary?', *History and Philosophy of the Life Sciences*, 21(3): 293–330.

Van Helden, A. (1983) 'The Birth of the Modern Scientific Instrument, 1550–1700', in J.G. Burke (ed) *The Uses of Science in the Age of Newton*, University of California Press, pp 49–84.

van Woerden, H. and Strom, R.G. (2006) 'The Beginnings of Radio Astronomy in the Netherlands', *Journal of Astronomical History and Heritage* 9(1): 3–20.

Varma, R. and Sabharwal, M. (2018) 'International Collaboration: Experiences of Indian Scientists and Engineers after Returning from the United States', *Perspectives on Global Development*, 17(5–6): 593–613.

von Hippel, E. (1976) 'The Dominant Role of Users in the Scientific Instrument Innovation Process', *Research Policy*, 5(3): 212–39.

Vonnegut, K. (1969) *Slaughterhouse-Five*, Seymour Lawrence/Delacorte.

Watt, I. (1957) *The Rise of the Novel*, Chatto and Windus.

Waugh, P. (1984) *Metafiction: The Theory and Practice of Self-Conscious Fiction*, Methuen.

Weber, W. (2003) *Music and the Middle Class* (2nd edn), Ashgate.

Weber, W. (2008) *The Great Transformation of Musical Taste*, Cambridge University Press.

Weingart, P. (2025) 'The Changed Publishing Culture of Science', *European Review*, 33(S1): 1–16.

Weston, J.F. (1974) 'New Themes in Finance', *Journal of Finance*, 29(1): 237–43.

White, H.C. and White, C. (1965) *Canvases and Careers: Institutional Change in the French Painting World*, Wiley.

Whitley, R. (1985) 'Knowledge Producers and Knowledge Acquirers: Popularisation as a Relation Between Scientific Fields and Their Publics', in T. Shinn and R. Whitley (eds) *Expository Science: Forms and Functions of Popularisation*, Reidel, pp 3–28.

Whitley, R. (1986a) 'The Rise of Modern Finance Theory: Its Characteristics as a Scientific Field and Connections to the Changing Structure of Capital Markets', *Research in the History of Economic Thought and Methodology* 4: 147–78.

Whitley, R. (1986b) 'The Transformation of Business Finance into Financial Economics: The Roles of Academic Expansion and changes in US capital Markets', *Accounting, Organizations and Society*, 11(2): 171–92.

Whitley, R. (2000 [1984]) *The Intellectual and Social Organization of the Sciences*, Oxford University Press.

Whitley, R. (2010) 'Reconfiguring the Public Sciences: The Impact of Governance Changes on Authority and Innovation in Public Science Systems' in R. Whitley, J. Gläser and L. Engwall (eds) *Reconfiguring Knowledge Production: Changing Authority Relationships in the Sciences and Their Consequences for Intellectual Innovation*, Oxford University Press, pp 3–47.

Whitley, R. (2014) 'How Do Institutional Changes Affect Scientific Innovations? The Effects of Shifts in Authority Relationships, Protected Space, and Flexibility', in R. Whitley and J. Gläser (eds) *Organizational Transformation and Scientific Change: The Impact of Institutional Restructuring on Universities and Intellectual Innovation*, Emerald Group, pp 367–406.

Whitley, R. and Gläser, J. (eds) (2007) *The Changing Governance of the Sciences: The Advent of Research Evaluation Systems*, Springer.

Whitley, R. and Gläser J. (2014) 'Editor's Introduction', in R. Whitley and J. Gläser (eds) *Organizational Transformation and Scientific Change: The Impact of Institutional Restructuring on Universities and Intellectual Innovation*, Emerald Group, pp 1–15.

Whitley, R., Gläser, J. and Laudel, G. (2018) 'The Impact of Changing Funding and Authority Relationships on Scientific Innovations', *Minerva*, 56(1): 109–34.

Whittall, A. (2009) '1909 and After: High Modernism and "New Music"', *Musical Times*, 150(1906): 5–18.

Wolff, J. (1993) *The Social Production of Art* (2nd edn), Macmillan.

Wondimagegn, D., Ragab, L., Yifter, H., Wassim, M., Rashid, M.A., Whitehead, C.R. et al (2022) 'Breaking Borders: How Barriers to Global Mobility Hinder International Partnerships in Academic Medicine', *Academic Medicine*, 97(1): 37–40.

Woodmansee, M. (1994) *The Author, Art, and the Market*, Columbia University Press.

Zapp, M. and Powell, J.J.W. (2016) 'How to Construct an Organizational Field: Empirical Educational Research in Germany, 1995–2015', *European Educational Research Journal*, 15(5): 537–57.

Zietsma, C., Groenewegen, P., Logue, D.M. and Hinings, C.R. (2016) 'Field or Fields? Building the Scaffolding for Cumulation of Research on Institutional Fields', *Academy of Management Annals*, 11(1): 391–450.

Zilsel, E. (1941) 'The Origins of William Gilbert's Scientific Method', *Journal of the History of Ideas*, 2(1): 1–32.

Zilsel, E. (1942) 'The Sociological Roots of Science', *American Journal of Sociology*, 47(4): 544–62.

Zilsel, E. (1945) 'The Genesis of the Concept of Scientific Progress', *Journal of the History of Ideas*, 6(3): 325–49.

Index

References in **bold** type refer to tables. References to endnotes show both the page number and the note number (171n1).

R

S

T

U